WORKPLACE INDUSTRIAL RELATIONS AND TECHNICAL CHANGE

WORKPLACE INDUSTRIAL RELATIONS AND TECHNICAL CHANGE

W. W. Daniel

Frances Pinter (Publishers)
in association with
Policy Studies Institute

First printed 1987
Reprinted 1987

British Library Cataloguing in Publication Data

Daniel, W. W.
Workplace industrial relations and technical change.
1. Industrial relations—Great Britain—Effect of technological innovations on
I. Title
331'.0941 HD8391

ISBN 0-86187-917-1
ISBN 0-86187-941-4 (PBK)

Printed by Blackmore Press, Longmead, Shaftesbury, Dorset

For Will, Bex and Sue

Contents

Foreword

Whilst many factors influence the nature of relationships between employees and their employers, it is increasingly recognised that systematic information is an important component for their development, both at the level of the individual workplace and at the national level. In the past much of the discussion and debate about industrial relations was based upon fragmented information and partial analysis. The first systematic survey of employment relations issues was undertaken for the Donovan Commission. Two other national surveys followed shortly afterwards. But these surveys never became part of a systematic series.

The idea of establishing a series of Workplace Industrial Relations Surveys was developed in the late 1970s in the Department of Employment to remedy this lack of systematic data and to make possible the analysis of change and continuity over time. The Economic and Social Research Council and the Policy Studies Institute, with the support of the Leverhulme Trust, felt that such a national series would provide unique and invaluable information for the analysis of both research and policy issues. As a consequence these organisations not only made a substantial financial contribution to the first survey but also made important inputs to the design, content and analyses.

The publication of the report on the first survey had a very favourable reception from commentators in the serious press and in journals concerned with the study and practice of industrial relations. One reviewer called it the most important book in the area since the publication of the Donovan report. In addition, however, to the initial analysis, the survey has formed the basis for studies of a wide range of topics including part-time working, trade union mark-ups, joint consultation, outworking, share ownership, personnel management, and pay structures. It has also generated a variety of linked studies on employee relations at the enterprise level and on the methodology of workplace surveys. Many of these studies were funded by the ESRC.

The reception given to the first survey and the many uses and publications which have resulted from it encouraged the continuation and development of the series. The second survey was carried out in 1984 with funding from the original three sponsors and also from the Advisory, Conciliation and Arbitration Service. The design for the second survey was in all essential respects similar to the first survey in order to provide a sound basis for the measurement of change. In terms of content, a substantial core of common questions was retained but a new topic area, concerned with technical change, was introduced into the survey and two other topics, employment practices and pay determination, were dealt with in more detail than in the first survey. In addition, an experimental panel element was built into the design.

The Second Workplace Industrial Relations Survey has been carried out under the guidance of a Steering Committee, which I have had the privilege of chairing, composed of representatives of the sponsoring bodies. The members of this committee were Professor George Bain, Francis Butler, Chris Caswill, John Cullinane, Bill Daniel, and Bill Hawes. I am grateful to them, not only for making an important contribution to the survey, but also for the tolerance they have shown for each others' differing interests in the enterprise and for their determination to co-operate for the greater good. The major responsibility for the design, execution and analysis of the research has, however, rested with the research team: Neil Millward and Mark Stevens from the Department of Employment, Bill Daniel from the Policy Studies Institute and Colin Airey of Social and Community Planning Research (SCPR). They have shown a determination in undertaking the work and a commitment to the series which has gone well beyond the bounds of duty or contract. It has been for me a pleasure to work with those who also share a belief that information is an essential component for rational discussion and analysis of employment matters.

There are two initial publications setting out the results of the second survey: one by Neil Millward and Mark Stevens which provides a general overview of the survey data; and this volume by W.W. Daniel which presents analysis of the material on new technology. The introduction of a section of questioning on the use and introduction of new technology into the survey has allowed the author to undertake a wide-ranging analysis of patterns of variation in the organisational and industrial relations processes surrounding technical change. Bill Daniel is able to compare the reactions of different categories of employees and managers in different types of workplace to a variety of forms of technical change. He is also able to disentangle reactions at different stages of the change process and to consider the implications for job content, work organisation and pay and employment. Whilst there have been many accounts of these issues based on small-scale studies, this is the first time that a comparative analysis across the economy has been possible. Daniel's account will, I am sure, become a benchmark for future discussions and analyses of the industrial relations implications of the use and adoption of new technology.

Besides the publication of the two general volumes and further articles that their authors are producing, a number of other analyses of the data are being undertaken with funding from the ESRC; these include work on management organisation, employment practices, wage levels, and labour relations and economic performance. In addition, there will be a special analysis of the panel data. As with the first survey, the data are deposited at the ESRC Data Archive and will, I am sure, provide an important source of information for both researchers and practitioners over future years.

The hope is that the series will continue so that comprehensive data will be available to policy makers, policy analysts and researchers on the changing and developing patterns of employment relations and associated issues across the British economy. Whilst such an enterprise is costly, its value seems to me to have been amply demonstrated. I have no doubt that readers of this volume and users of the data on which it is based will agree.

Peter Brannen
Chairman, Steering Committee for the Workplace Industrial Relations Surveys

Author's Acknowledgements

When PSI produced the draft of its triennial report in 1983 and circulated it to members of Council for comment, one member complained that we used the word *unique* to describe aspects of projects on three separate occasions. That, he suggested, represented unnecessary *hype*. We should be more modest in the claims for our work. He allowed, however, that we had every justification for using the word in one context; that where we referred to the *unique collaboration between a Government Department, a national Research Council and an independent Research Institute to establish the Workplace Industrial Relations Survey Series*. PSI regards its involvement in the series as special and is grateful to the Department of Employment (DE), the Economic and Social Research Council (ESRC) and latterly the Advisory Conciliation and Arbitration Service (ACAS) for collaborating with us in the enterprise. Our involvement in both the 1980 and 1984 surveys was made possible by generous support from the Leverhulme Trust.

The present volume which we have produced from the 1984 survey owes much to two programmes of research in progress at PSI. The first is concerned with industrial relations and the second with the diffusion of new technology. The outcome is evidence of the benefit derived from having streams of work on different topics being carried out under the same roof. I have been responsible for the industrial relations programme. Jim Northcott set up and directs the new technology programme. The output from that programme was invaluable to me in this work and is referred to repeatedly throughout the book. The establishment of PSI's industrial relations and new technology programmes owed much to both the Leverhulme Trust and the Joseph Rowntree Memorial Trust (JRMT). Karen McKinnon carried out the computer analysis to produce the tables from the survey on which the book is based. Her skill and familiarity with the Quantum system of analysis which we use at PSI meant that that analysis was produced with a speed and accuracy that I have not previously experienced. Eileen Reid and Margaret Cornell edited the text.

Finally, in thanking all those individuals and organisations mentioned I would like to add my thanks to Peter Brannen, Chairman of the Steering Group whose inspiration created the show and who keeps it on the road;

to other members of the Steering Committee, the Research Team and staff at Social and Community Planning Research (SCPR) that he mentions in the *Foreword*; and to the managers and shop stewards who took part in the survey.

Notes on tables used in the text

General conventions adopted in tables

- § Unweighted base too low for percentages (see Note A).
- ( ) Percentages should be treated with caution (see Note B).
- ⋆ Fewer than 0.5 per cent
- – Zero

Notes on tables

(A) Unweighted base is fewer than 20 and therefore too low for percentages

(B) Unweighted base is 20 or more but fewer than 50; percentages should be treated with caution

(C) Column and row percentages do not always add to 100 owing to the rounding of decimal points

(D) The proportions in subsidiary categories do not always add to the proportion in a composite category owing to the rounding of decimal points

(E) Column and row percentages sometimes add to more than 100 because more than one answer was possible

(F) The proportions in subsidiary categories do not always add to the proportion in a composite category because more than one answer was possible

(G) The base numbers for the individual categories in a variable do not always add to the total base number because the necessary information was not provided in a number of cases

(H) The base numbers for the individual categories in a variable do not add to the total base number because only an illustrative range of categories is included in the table

(I) In cases where the answers of respondents were recorded on a five point scale or a three point scale we generally calculated and show a *mean score* in tables. An example is provided by accounts of workers' reactions to change where answers varied from *strong support* at one extreme to *strong resistance* at the other. The maximum score was +200 which would have been achieved if all respondents said that there was *strong support*. A score of +100 represented the equivalent of all respondents saying that there was *slight support*. The scores were calculated by giving an arbitrary value of +200 to *strong support*, +100 to *slight support*, -100 to *slight resistance* and -200 to *strong resistance* and dividing the sum by the proportion who expressed views. The result was then rounded to the nearest ten to provide a mean score.

The different sources of information

As in the case with the text, where information is presented in tables without its source being identified it will have been provided by the principal management respondent.

Unless otherwise specified the information used for breakdown variables, such as number employed at the establishment, size of total organisation, industrial sector, ownership and trade union membership density, will also be that provided by principal management respondents.

The All Establishments and All Employees columns

The weighted sample upon which percentages are based is representative of establishments in the economy employing 25 or more people. As a large proportion of the nation's work force is employed at a small proportion of its establishments we often include in tables an *All Employees* column which shows what proportion of employees at the establishments included in the table fall into the categories reported (see *Introduction*).

I Introduction

When establishing the workplace industrial relations series, we always had two main purposes in mind(1). The first was to provide a basis for monitoring change. For that purpose, we adopted a common design for each survey in the series and included a number of core questions which were selected to establish the nature and extent of change over time in a number of key industrial relations characteristics of workplaces. The core coverage included practices, patterns and issues relating to trade union membership and trade union recognition; trade union organisation at the workplace; the closed shop; systems of pay determination; industrial relations procedures; industrial action; consultative committees; and arrangements for managing industrial relations. The companion volume published in parallel with this book analyses the changes that occurred between 1980 and 1984, according to answers to the core questions asked in the first two surveys in our series.

Secondly, we had in mind that each survey would address one or more new substantive subjects of topical interest and importance for industrial relations practice and policy. The scale of the surveys in the workplace series, the comprehensiveness of their coverage and the inclusion of interviews with manual and non-manual shop stewards as well as managers would enable us to address such substantive issues more authoritatively than is often possible with the resources available for particular, *ad hoc* surveys. Moreover, the thorough coverage of industrial relations arrangements, institutions, practices and procedures provided by the core questions in the survey would give us an unrivalled opportunity to analyse matters relating to the new substantive subject in relation to such characteristics. In the 1984 workplace survey we adopted the introduction of change, and particularly change involving new technology, as the main new substantive subject. This volume represents our analysis of technical change on the basis of the results of that survey.

The coverage of the survey relating to technical change

We set out to establish how far the industrial relations arrangements at workplaces were associated with their use of the most modern technology and the manner in which it was introduced. The core questioning in the survey gave us unrivalled information about labour relations institutions, practices and procedures including trade union representation, recognition and densities, multi-unionism, the closed shop and bargaining and consultative structures. Moreover, we had those details for both major groups of employee: non-manual workers as well as manual workers. We decided to focus upon the adoption of microelectronic technology as our measure of their use of new technology. Microelectronic technology became generally available in the second half of the 1970s. The nature of the technology may best be summarised by quoting the definition that we provided where necessary for respondents in our survey.

> *The use of the new microelectronics technology* is defined as *the use of microprocessors or their equivalent electronic alternatives – including single devices (chips) or small linked groups of devices or both*. The device will normally involve *large-scale integration (LSI)* or *very large-scale integration (VLSI)* and be used in combination with other kinds of integrated circuits. The use of microelectronics technology includes the use of computer *controlled* plant, machinery or equipment.

The technology was widely hailed as a massive new source of power that would transform processes both in the workshop and in the office. People talked of a second industrial revolution. By the early 1980s we at PSI had substantial experience of plotting the introduction of microelectronic applications in manufacturing industry(2). In 1983 we had found that 43 per cent of factories were using processes that incorporated microelectronics and 10 per cent were making products that included the new technology. Usage was most common in the engineering industries, and in process industries (see Figure I.1). Microelectronics were being used in well over one hundred different types of product. Indeed, they tended to be used in almost any product where there was something to be measured, monitored or controlled. So far as process applications were concerned, microelectronics were most frequently used in the control of individual machines or processes but their application to the integrated central control of groups of machines or processes was beginning to grow. Types of process equipment used included programmable logic controllers (PLCs); computer numerical controlled (CNC) machine tools; pick-and-place machines; and robots (see Figure I.2). So far as office applications were concerned, PSI found that in 1983 two thirds of manufacturing establishments were using computer facilities and nearly one fifth were using word processors.

This picture persuaded us that if we included in the 1984 workplace industrial relations survey (WIRS 84) questioning about microelectronic applications we would have a good measure of the modernity and innovativeness of the workplace which we could relate to its labour relations arrangements. Our previous experience of researching the introduction of new technology enabled us quickly to develop a set of questions to serve as that measure. Their inclusion in WIRS 1984 made it possible to extend the account and analysis of the use of new technology beyond

Figure 1.1
USE OF MICROELECTRONICS BY INDUSTRY

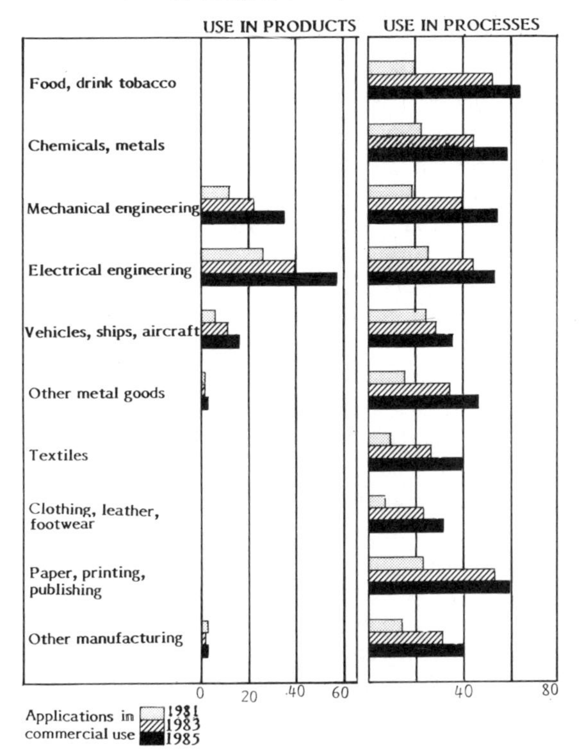

Figure 1.2
TYPE OF PROCESS EQUIPMENT USED

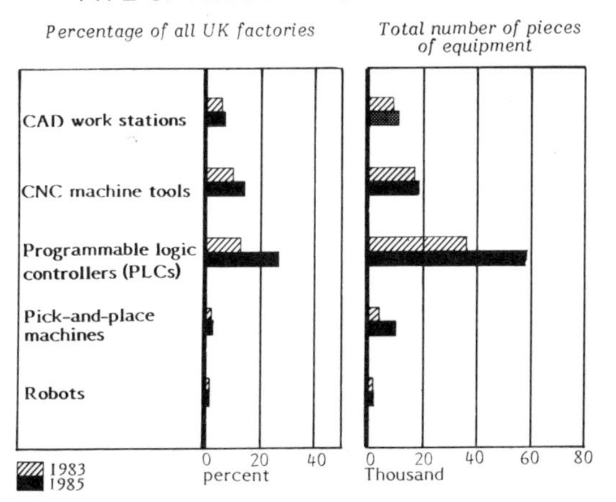

private manufacturing to private and public services. That extension also made it especially appropriate to devote as much attention to the use of new technology in the office as to its use on the shop floor.

The questioning on new technology in WIRS 1984 is reproduced in *Appendix A* and took three forms. The questions asked about applications affecting manual workers differed slightly from the equivalents relating to non-manual workers. So far as manual workers were concerned, we sought, first, to establish whether workplaces were using any application of microelectronics that directly affected the jobs of manual workers. In manufacturing industry, we distinguished between product and process applications. When asking the questions, interviewers briefed managers and shop stewards, if necessary, with cards carrying the definition of microelectronics given above.

Secondly, we wanted to find out how far changes involving the new technology differed from other types of change to which workers have been subject for generations. We assumed that there would be little inherent difference. As we explain more fully in Chapter IV, we expected that the way in which change was received would depend chiefly upon its implications for earnings, employment and the intrinsic interest of work. In addition, we had good reason to suppose that the manner of its introduction would prove to be important, especially the extent to which the workers affected and their representatives were involved in decisions about the change. Insofar as the public debate about the microelectronics revolution often seemed to be predicated upon the idea that the introduction of the new technology constituted a quite distinctive form of change, it seemed important for public policy-makers, managers and trade union representatives to know whether this was so and in what ways it was distinctive. For that purpose, we distinguished between three forms of change which we subsequently labelled (a) *advanced technical change*; (b) *conventional technical change* and (c) *organisational change*. We defined each as follows:

Advanced technical change

The introduction of new plant, machinery or equipment, that *includes* the new microelectronics technology (including computer controlled plant, machinery or equipment);

Conventional technical change

The introduction of new plant, machinery or equipment, *not* incorporating microelectronics (excluding routine replacement);

Organisational change

Substantial changes in work organisation or working practices *not* involving new plant, machinery or equipment.

It may seem a little odd in 1986 to be talking of any changes involving the use of microelectronics as *advanced* technical change. Indeed, engineers tend to confine the use of that term to the newest and most sophisticated forms of new technology. But, in reporting the results of our study, it is necessary for us to distinguish between technical change which involved microelectronics and that which did not, without using such clumsy phrases throughout, and the adjectives *advanced* and *conventional*, respectively, appeared the most appropriate way of doing so. When in the report we talk simply of *technical change* we are referring to both forms. Occasionally, in the report we talk about *the new technology* or *advanced technology* when referring to technology incorporating microelectronics, in order to provide stylistic variation.

Our distinction between the three different forms of change enabled us to examine how far the impact of *advanced technical change* differed from that of other forms of change. In the event we found that there were marked contrasts in reactions to the three forms of change and that the distinction served as one of the major sources of variation in our analysis. The second purpose of the distinction was that it enabled us to identify specific changes of each of the three types which subsequently served as the focus for our third area of questioning about the impact and implications of change. For instance, having established that there was an *advanced technical change* at the workplace we identified the most recent change of that type. We then asked about the level in the organisation at which the decision to introduce that particular change was taken; whether there were consultations with external advisers about the introduction of the change; what, if any, discussions were held with workers or their representatives about the change; what impact the change had upon earnings, manning and job content in the section of manual workers directly affected; and what the reactions to the change were from first-line managers, the workers affected, shop stewards and any full-time union officers involved. We emphasise that each of these questions focussed upon particular, named changes because it was clear from our analysis that we established a better guide to the normal behaviour of managers when we asked them what they did in particular cases rather than asking them what they did generally.

When we came to look at the impact of advanced technology upon non-manual workers, it seemed both desirable and possible to focus upon a category that was common to all types of workplace, in order that we might be able to make comparisons acoss sectors. Office workers, whom we defined as *clerical, secretarial, administrative and typing staff*, appeared to represent a group appropriate for such standardisation. The chief implication of the development of microelectronics for office workers was that it made possible a new generation of relatively low cost mini and micro computers and word processors. Accordingly, the focus for our questioning about advanced technical change affecting non-manual workers was the introduction of computers and word processors in the office.

The survey design

The design of the 1984 survey followed closely that adopted in 1980 and had the same three key features(3). First, we made the establishment our principal unit of analysis. Secondly, we carried out interviews with both management and worker representatives at workplaces. Thirdly, the coverage was comprehensive.

By *establishment* we mean an individual place of employment at a single address. The large majority of such places of employment in the economy belong to larger organisations which operate a number of workplaces. In our survey, 82 per cent of the sample of establishments were parts of larger organisations. In public services and nationalised industries that was almost invariably the case. In private services it was true for 74 per cent of establishments; and in private manufacturing for 68 per cent. Critics of workplace surveys have sometimes suggested that they give an exaggerated impression of the autonomy of local managers and trade union representatives in industrial relations matters and fail to show the extent to which local choices and practices are constrained by centralised policies at head office or intermediate levels. The answer to that criticism, of course, is that it is possible to do only so much in one survey. The design of the surveys in this series is more comprehensive and ambitious than any of its type previously attempted. It is appropriate to carry out workplace surveys in order to establish the features of industrial relations that can be identified at only that level. Other enquiries at other levels are necessary to establish what can be learned from only head offices, or divisional offices or regional offices. At the same time, we did wherever possible in our interviews with local managers and shop stewards seek to establish how far decisions about particular topics were taken at higher levels. In order to provide stylistic variation we refer interchangeably to establishments in the report as *workplaces, places, sites, units* or *establishments*, and when talking exclusively about manufacturing establishments we often refer to *plants*, *workshops* or *factories*. The larger organisations of which establishments are part we refer to in a similarly interchangeable way as *enterprises* or *organisations*. When the focus is exclusively on the private sector we refer to *businesses, companies* or *firms*.

One of the chief problems in designing and writing about surveys of workplaces is that a small proportion of establishments employ a large proportion of the working population. For example, large workplaces (those employing 1,000 or more people) made up only one per cent of the establishments in our survey but they accounted for 19 per cent of the workers covered by the survey. Where a characteristic of a workplace was strongly associated with its size then the proportion of establishments having the characteristic was often very different from the proportion of workers covered by it. For instance, 15 per cent of workplaces introduced *advanced technical change* affecting manual workers in the three years leading up to our interviewing, but 33 per cent of manual workers were employed at those places. Our sample of 2,000 workplaces was carefully designed in order to enable us to say both what proportion of workplaces operated a particular practice and what proportion of employees was covered by the practice. In tables we often present two *total* columns, one giving percentages based on *all establishments* and the other based on *all employees*.

The second key feature of the survey design was multiple interviews with different types of manager and with shop stewards. We describe the nature of our respondents when reporting on the sample and response rate in the next section. We discuss at the end of our introduction the great benefit of having such multiple interviews when reviewing the special strengths of the workplace series for analysing new substantive topics. The third feature of the design was its comprehensiveness. As in 1980, we covered all parts of the manufacturing and service sectors of civil employment in Great Britain(4). We were able to compare private manufacturing, private services, public services and nationalised industries. That served as a great advantage for the subject matter of this book, as no systematic information was previously available on the extent to which advanced technology was used or the extent of technical and other forms of change outside private manufacturing. In order to make the comparisons across sectors, we carried out interviewing at just over 2,000 workplaces.

As the principal interest of the series lies in institutional industrial relations, we did not include very small workplaces in the survey. We excluded, however, only those places that employed fewer than 25 people, a cut-off point substantially lower than that generally used in surveys of establishments concerned with industrial relations.

The sample and overall response

The sample of establishments was drawn from the 1981 *Census of Employment*, the most comprehensive and up-to-date listing of employment units available. A random, unclustered sample of census units was drawn, stratified in relation to the number of people employed. Interviews that satisfied our rigorous criteria for inclusion were completed at 2,019 workplaces which represented a response rate of 77 per cent. That overall response rate was calculated on the basis of the number of establishments at which an interview was completed with a management respondent and details of the composition of the workforce were collected. In addition, we also sought interviews with shop stewards and works managers in appropriate cases.

The respondents
Initially, interviewers were asked to identify and interview the senior person at the establishment with responsibility for industrial relations, employee relations or personnel matters. In fact, as we discuss in our companion volume, it was only in a minority of cases that those *principal management respondents* were personnel or industrial relations managers. More frequently, they were general managers such as *branch managers* or *office managers*. Subsequently, in places where manual unions were recognised, we sought through managers to contact and interview the senior shop steward representing the largest manual group at the workplace. Similarly, in places where non-manual unions or staff associations were recognised we set out to interview the senior representative of the largest group of non-manual workers. Finally, in manufacturing plants where our principal management respondent was a personnel manager we sought also to interview the works or production manager. Consequently, in larger manufacturing plants, we often ended up with four interviews; with the personnel manager, the works manager, the senior manual shop steward and the senior non-manual shop steward. In contrast, a small private service establishment would typically yield only one interview with the branch manager. We interviewed manual shop stewards in 79 per cent of cases where we sought them, and non-manual stewards in 82 per cent of cases. We were successful in talking to works managers in 74 per cent of places where it was appropriate to do so. Figure I.3 provides a summary of the different types of respondent that we interviewed.

In instances where we had more than one interview, we often asked the same or a very similar question of two or more people. As we had up to four interviews at each workplace, we often had a choice as to whose information we should present as reflecting a particular characteristic of the workplace. Generally, we use the interviews with our primary management respondent to provide the widest coverage. We used the interviews with stewards and works managers for comparative purposes and to supplement the results of the main management interview where different or additional questions were asked.

Interviews were carried out over the period of mid-March to early October, 1984, and about three quarters of them were completed during the four months April to July. Interviews with principal management respondents typically lasted one and a half hours, while those with shop stewards and works managers took about three quarters of an hour.

Figure I.3 *Summary of number of respondents*

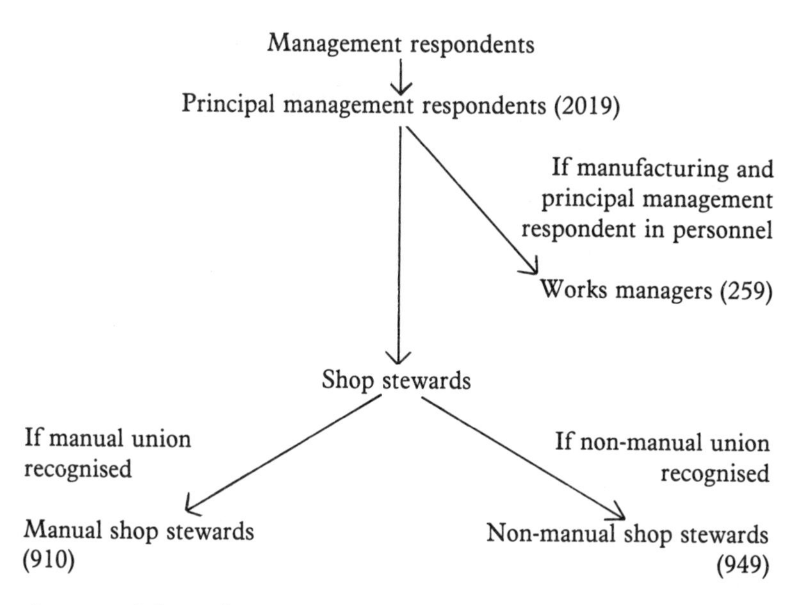

Accuracy of the results

Because we oversampled larger establishments in the way described in the next section, sampling error was slightly greater than would have been the case had simple random sampling been used. Table I.1 shows the approximate sampling error for a selection of proportions for the whole sample survey, with a 95 per cent level of confidence.

Table I.1 Sampling errors

	Percentage found by survey				
	5 or 95	10 or 90	20 or 80	30 or 70	50
Approximate sampling error	1.3	1.6	2.3	2.5	2.8

Accordingly, when we found, for instance, that 15 per cent of workplaces introduced advanced technical change affecting manual workers in the three years previous to the survey, we could be confident, given that there was no other source of error, that that was the case for between 13 per cent and 17 per cent of workplaces.

Units of analysis and weighting

As larger workplaces represent such a small proportion of all establishments but account for such a large proportion of all employees, and as we wanted to ensure that we had a good basis for analysing variations in our results in relation to size, we elected to oversample larger workplaces. For that purpose, the initial sample from the census was stratified according to size of establishment and different sampling fractions applied to each size band. Subsequently, weights were attached to each size band in order to restore the numbers to their proper proportions according to the census. For convenience, the results have been weighted to a base of 2,000 establishments. Further details of the weighting together with the fieldwork and other methodological aspects of the survey are summarised in Appendix B and reported more fully in the Technical Appendix to our companion volume(5).

We have used detailed tabular analysis to explore the results of the survey for the purposes of this book. That method of analysis was adopted because it provided a fast and flexible way of identifying the aspects of workplaces that were associated with particular variables, such as the use of new technology, and presenting them in a manner that is understandable to a general readership. In fact, we found as the analysis proceeded that it tended to fall into a pattern. First, we looked at the results in relation to the two main structural characteristics of workplaces that tended to be major sources of variation throughout. They were *sector* and *size*. In relation to sector, we distinguished between private manufacturing industry, private services, nationalised industries and public services. In terms of size, we focussed principally upon the number of people employed at the workplace, though the total size of the organisation of which it was part was often also of importance. When the question referred to manual workers, we tended to use the number of manual workers employed as the measure of establishment size and when the question referred to non-manual workers or office workers we adopted the number of non-manual workers at the establishment as the measure of size.

Secondly, in the private sector, we focussed upon differences between workplaces in different industrial sectors, between domestically-owned workplaces and those that belonged to overseas companies, and between independent establishments and those that were parts of larger organisations. In particular, we found throughout that the nationality of the ownership of workplaces was a major source of variation in their use of new technology. And, as we highlight in our conclusions, the contrast was not favourable to British companies.

Thirdly, we looked at any associations between the particular topic and the trade union arrangements at the workplace, taking into account other major sources of variation that might already have emerged. For that purpose, we found the distinction between places that recognised unions and those that did not, and variations between those with differing levels of *trade union density*, to be the most fruitful forms of analysis. In each instance, we distinguished between the recognition of manual and of non-manual unions and between manual and non-manual union densities. Here we should emphasise that this sequential form of tabular analysis did tend to follow from the variations in our data.

Special strengths of the workplace series for analysing new substantive topics

Three main characteristics distinguish surveys in the WIRS series from their predecessors. They cover all sectors rather than concentrating upon manufacturing industry. They devote as much space to non-manual workers as to manual grades. They include interviews with both manual and non-manual shop stewards. We found that these distinguishing characteristics were even greater sources of strength in the present analysis of technical change as a new substantive topic than they were in the analysis of basic industrial relations institutions. First, the contrasts between private manufacturing and nationalised industries and between private services and public services were a recurrent source of variation throughout the report and provide an important plank for the conclusions. Secondly, the coverage of non-manual workers revealed that technical change has in the recent past become more common for office workers than for manual workers. Moreover, the extent to which computers and word processors were used in offices provided a standardised measure across sectors of the modernity of workplaces that was not available at the manual level.

Thirdly, the multiple interviewing at workplaces proved invaluable in this study. Some commentators have criticised the interviews with shop stewards as well as managers in the workplace survey as an unnecessary luxury(6). For the purposes of our present analysis they were indispensable. For instance, the pictures revealed by our questioning about consultations over the introduction of change and about the respective reactions of the workers affected and their union representatives to proposed changes were substantially different when based upon management accounts from those based upon union accounts. Indeed, so substantial were the differences that it would be difficult, in future, to attach much weight to information on such matters that was based upon interviewing with only one of the parties. The types of question that most required interviews with both sides of industry were those which might appear to be fairly straightforward matters of fact but which, in practice, were susceptible to genuine differences in the way that different parties saw or experienced them owing to their roles and relationships. For instance, there are good reasons why, in some circumstances, shop stewards might give the impression that they were less strongly in favour of particular changes than their members when, in fact, stewards themselves judged that they were more strongly in favour. Similarly, it is clear that it is perfectly possible for stewards to feel that they have not been consulted over particular changes when managers feel that they have been consulted. It was in relation to questions that were at an intermediate point between simple, straightforward factual matters at one extreme and attitudinal matters at the other that our multiple interviewing was of most value. The parts of our interviews concerned with the introduction of change included a larger proportion of such intermediate questions than the core questioning in the survey which was closer to the more straightforwardly factual end of the spectrum. In addition, of course, at a simple, practical level it was a great advantage to be able to include in one set of interviews questions which we did not have sufficient space to ask in another. For example, we were able to distinguish in our interviews with shop stewards between consultations over *whether* to introduce the change and consultations over *how* to introduce it. Similarly, we were able to distinguish between the reactions of workers and shop stewards to initial news of the proposed change compared with their feelings after it was implemented. Both distinctions substantially enriched our analysis.

In addition, our interviews with works managers in those parts of manufacturing industry where our principal management respondents were personnel managers further enriched the analysis. In particular, those interviews provided us with a very useful safeguard against the possibility that our information at a number of points might be coloured by a personnel management perspective. Furthermore, as with our shop steward interviews, we were able to use the works manager schedule to collect valuable additional information. In particular, our analysis in Chapter VII of the association between the introduction of *advanced technology* and the increasing flexibility of working practices on the shop floor was based upon the interviews with works managers and very much strengthened our analysis of the introduction of technical change. Also, the evidence of the value of involving personnel managers in the introduction of change was strengthened by the fact that it came from works managers.

Footnotes

(1) The first survey in the series was carried out in 1980 and published as W.W. Daniel and Neil Millward, *Workplace Industrial Relations in Britain*, Heinemann, London, 1983. This book is based upon the second survey in the series, carried out in 1984 and published in this volume and Neil Millward and Mark Stevens, *British Workplace Industrial Relations - 1980 to 1984*, Gower, Aldershot, 1986.

(2) Since the late 1970s Jim Northcott has been directing a programme of research at PSI on the introduction of new technology involving microelectronics. By 1984 that had involved two major surveys of usage in manufacturing industry, Jim Northcott with Petra Rogers, *Microelectronics in Industry: What's Happening in Britain*, PSI, No.603,

1982 and Jim Northcott and Petra Rogers, *Microelectronics in British Industry: The Pattern of Change*, PSI, No.625, 1984. Subsequently a third survey was completed in 1985, Jim Northcott, *Microelectronics in Industry: Promise and Performance*, PSI, No.657, 1986.

(3) Daniel and Millward, 1983, *op.cit.*

(4) Primary industries (agriculture, forestry and fishing) were not covered as they are not included in the census of employment. In 1980 we did not manage to cover coalmining because negotiations with the industry were not completed in time for it to be included. That left us with the dilemma as to whether to seek to include coalmining in 1984 thereby making the survey more comprehensive but complicating the processes of comparison between the two surveys. The dispute in the industry over the period of the survey resolved that dilemma for us.

(5) Millward and Stevens, 1986, *op.cit.*

(6) Submissions to Economic and Social Research Council (ESRC) on the design of the second survey in the series, following analysis of the first.

II Change and the Use of Microelectronics on the Shop Floor

'Change is the only constant', declaimed Graham Day, newly appointed Chairman of British Leyland, to Brian Redhead as I drove to the office to start this chapter(1). He could have been any senior manager on any morning during the past quarter of a century, promoting one of the slogans of his occupation. Our present findings on the extent of major change at work, however, suggest that reality has very nearly caught up with the rhetoric. Not only does the idea of the permanence of change accord with the purposes of management and hence its ideology but it has also come to reflect what is happening on the ground. The subject of our study was the introduction and use of *advanced technology*. But in order to put that subject into perspective, we also wanted to establish the incidence of other forms of change. Accordingly, we distinguished between *advanced technical change*, *conventional technical change*, and *organisational change*, as defined on page 4 in Chapter 1. As well as asking managers and shop stewards about their use of different microelectronic applications, we also asked them if they had experienced changes of each of these three types. The question proved to be a key one as we show in Chapter IV, where we discuss the contrasting reactions to the three different forms of change. Here, in the first part of the chapter we confine ourselves to an account of their frequency and distribution, as they affected manual workers. We subsequently go on to a more detailed analysis of the distribution of microelectronic applications in relation to manual trade union organisation.

The extent of major changes affecting manual workers

In summary, answers to our question on the three forms of change revealed that the majority of manual workers were employed at workplaces that had introduced major change in the recent past. The most common form of innovation consisted of changes in the organisation of work or working practices, not involving the introduction of new machines or equipment. Organisational change, however, was only slightly more common than conventional technical change which, in turn, was only slightly more common than advanced technical change. Moreover, as technical change, in general, and advanced technical change, in particular, were more common in larger workplaces, they tended to affect more workers. Workplaces that recognised trade unions were more likely to have introduced major changes affecting manual workers than those that recognised no manual union.

Table II.1 The extent of major change affecting manual workers in the previous three years

Percentages

	All sectors	Manufacturing industries
Any of the three forms of change		
Proportion of workplaces experiencing change	37	65
Proportion of manual workers employed at workplaces experiencing change	57	78
Proportion of manual workers at establishment covered by most recent specific change	36	27
Advanced technical change		
Proportion of workplaces experiencing change	15	31
Proportion of manual workers employed at workplaces experiencing change	33	53
Proportion of manual workers at establishment covered by most recent specific change	26	24

At least one of the three forms of change affecting manual workers had been introduced into over one third of workplaces (see Table II.1). As change was increasingly more common the larger the number of manual workers employed (as shown in Table II.4), the majority of manual workers (57 per cent) were employed at workplaces recently experiencing substantial change affecting manual grades. After we identified the most recent major change introduced at workplaces we asked how many manual workers were affected by the change. We found that, in general, the coverage was just over one third of the manual workforce. From that proportion and the number of manual workers employed at establishments where change had occurred, we were able to calculate that 16 per cent of all manual workers were personally affected by the most recent change introduced at workplaces.

Consequently, it was clear that a maximum of 57 per cent and a minimum of 16 per cent of all manual workers were directly affected by one or

Table II.2 The extent to which establishments experienced more than one of the three forms of major change

Column percentages

	All establishments	Number of manual workers at establishment						All manual employees
		1-24	25-49	50-99	100-199	200-499	500 or more	
All three forms of change	3	*	3	4	8	14	27	12
Two of the three forms of change	8	4	9	14	16	18	23	15
One of the three forms of change	26	19	29	35	27	36	31	30
No changes	63	77	59	47	49	32	19	43
Base: establishments that employed manual workers								
Unweighted	*1853*	*430*	*349*	*255*	*259*	*301*	*259*	
Weighted	*1749*	*764*	*345*	*240*	*117*	*64*	*20*	

more of the three types of change we identified in the previous three years. The 57 per cent represented the proportion of all manual workers employed at establishments that introduced some change. The 16 per cent represented the proportion of all manual workers affected by the single most recent change introduced. It is likely that the actual proportion of manual workers affected by change was nearer the top than the bottom of that range. Many establishments experienced more than one of the three forms of change (see Tables II.2 and II.3). Of the large workplaces, one half experienced at least two of the three forms of change we identified, while over one quarter experienced all three. In addition, of course, some of the establishments that introduced a particular form of change, will have introduced more than one innovation of that type.

Table II.3 The overlap between the three different forms of change in the previous three years

Column percentages

	Total	Advanced technical change	Conventional technical change	Organisational change
Advanced technical change	39	100	31	24
Conventional technical change	48	39	100	37
Organisational change	51	31	39	100
Base: establishments experiencing some major change affecting manual workers				
Unweighted	*909*	*478*	*504*	*489*
Weighted	*647*	*255*	*313*	*329*

It is apparent from this analysis that major changes of the type we defined have become common for manual workers. The changes tended to be large in scale and scope. Many workers were directly affected. Changes often had an impact upon earnings, manning and job content (see Chapter IV). *Advanced technical change* formed a large part of the pattern. Although, in terms of the proportion of workplaces, *organisational change* was the most common of the three forms, and *conventional technical change* was the second most common, *advanced technical change* was most common at the larger workplaces. In consequence, a similar number of manual workers were employed at places introducing advanced technical change as at places introducing either of the other forms of change (see Table II.4). Indeed, one third of all manual workers were employed at workplaces that introduced advanced technical change

Table II.4 Types of change introduced in previous three years

Column percentages

	All establishments	Number of manual workers at establishment: 1-49	50-199	200 or more	All manual employees
Some major change	37	31	52	72	57
Advanced technical change	*15*[a]	*10*	*23*	*46*	*32*
Conventional technical change	*18*	*14*	*29*	*40*	*32*
Organisational change	*19*	*16*	*24*	*40*	*32*
No change	63	69	48	28	43
Proportion (per cent) of manual workers affected[b]	36	45	32	22	
Base: all establishments employing manual workers					
Unweighted	*1853*	*779*	*514*	*560*	
Weighted	*1749*	*1308*	*357*	*84*	

[a] See note F.

[b] This percentage represents the proportion of manual workers covered by the most recent change at establishments that experienced change.

Table II.5 Extent to which establishments experienced different forms of change within the previous three years

Column percentages[a]

	All establishments					Large establishments employing 500-999 employees			
	Total	Private manufacturing	Private services	Nationalised industries	Public services	Private manufacturing	Private services	Nationalised industries	Public services
Advanced technical change	15	31	8	25	8	71	(17)[b]	(41)	23
Conventional technical change	18	38	7	28	14	51	(28)	(27)	22
Organisational change	19	25	15	28	18	45	(18)	(50)	21
No change	63	35	76	48	69	13	(55)	(40)	44
Not stated	*	*	*	1	*	1	–	–	*
Base: all establishments employing manual workers									
Unweighted	*1853*	*580*	*515*	*191*	*567*	*118*	*43*	*38*	*87*
Weighted	*1749*	*412*	*689*	*103*	*545*	*14*	*6*	*5*	*12*

[a] See note E.
[b] See note B.

affecting the jobs of manual grades in the previous three years. In manufacturing industry, just over one half of manual workers were employed at such establishments.

Table II.5 shows, in a different way, the concentration in manufacturing industry of change affecting manual workers, in general, and change involving *advanced technology*, in particular. The introduction of both forms of new technology was very much more common in private manufacturing industry than in both private and public services. Only nationalised industries came close to private manufacturing in the introduction of new technology affecting manual jobs, and, of course, nationalised industries included a minority of manufacturing units. Organisational change affecting manual jobs was more common than technological change in both private and public services but even organisational change was less common than in private manufacturing industry. Nationalised industries, were, however, most affected by organisational change.

The pattern of variation revealed so far in the extent to which different types of workplace introduced particular changes involving advanced technology foreshadows the general distribution of microelectronic processes which we report later in the chapter. That is as would be expected. More surprisingly, the distribution of the other two forms of change appeared to follow similar lines. There were, however, important ways in which our picture of the introduction of different particular forms of change differed subtly from the later pattern showing the general distribution of microelectronics. For instance, Table II.4 shows that each of the three types of change was more common the larger the number of manual workers employed on site. But the order of frequency with which each occurred varied as between the different size bands. In the smallest workplaces, organisational change was most common and advanced technical change was least frequent. At medium sized workplaces, new conventional technology was the most common vehicle of change. In the largest workplaces the introduction of advanced technology was the most common form of change.

Two particular characteristics strongly associated with our measures of the introduction of specific changes in a way that was not related to the general distribution of microelectronics, were the labour intensity of workplaces and managers' assessment of the sensitivity of their product markets to price change. As might have been expected, technological innovation was, first, more common the higher the level of capital intensity. Secondly, the more competitive were product markets, according to managers' assessments, in terms of price, the more likely were workplaces to have introduced change. The trend is consistent with the possibility that increasing price competition does provide a stimulus to innovation.

Our later findings on the distribution of microelectronics show that overseas companies were way ahead of domestic counterparts in their use of the new technology. Table II.6 shows that not only were establishments belonging to overseas companies more likely than UK counterparts to have introduced advanced technology but they were also more active in the other forms of change. Establishments which belonged to a group were very much more likely than independent establishments to have been subject to organisational change not associated with the introduction of new technology. That contrast was especially strong in manufacturing industry.

Table II.6 Change in previous three years in relation to certain aspects of ownership

Column percentages[a]

	UK owned	Overseas owned	Independent establishment	Part of group
All establishments				
Advanced technical change	16	27	15	14
Conventional technical change	18	29	21	17
Organisational change	17	30	12	21
Any major change	38[b]	54	38	37
Base: all employing manual workers				
Unweighted	*927*	*168*	*192*	*1452*
Weighted	*1006*	*96*	*328*	*1287*
Manufacturing establishments				
Advanced technical change	29	48	26	34
Conventional technical change	37	46	40	39
Organisational change	24	29	14	30
Any major change	64	75	62	67
Base: manufacturing workplaces employing manual workers				
Unweighted	*468*	*113*	*80*	*438*
Weighted	*367*	*452*	*137*	*238*

[a] See note E.
[b] See note F.

The introduction of change in relation to manual trade union organisation at the workplace

As we mentioned at the beginning of the chapter, we found clear evidence that workplaces which recognised manual trade unions were consistently more likely to introduce change affecting manual workers than establishments which did not. That tendency persisted independently of differences in the size and sector of plants. When, however, we looked for any association between manual trade union density and the introduction of change, we found none. The pattern was very mixed and trends tended to move in different directions in different sectors and size bands.

Change affecting manual workers tended to be concentrated in manufacturing industry and among larger establishments. Similarly, trade union organisation among manual workers tended to be greatest in manufacturing and at large workplaces. In consequence, it was no surprise, overall, to find that both the introduction of advanced technology and other forms of change tended to be substantially more common in workplaces where manual unions were recognised than in those where they were not (see Table II.7). Forty per cent of establishments which recognised manual unions had introduced substantial change affecting manual workers in the previous three years compared with 33 per cent of establishments where manual unions were not recognised. When comparisons were made between establishments employing similar numbers of manual workers, it remained the case that there was a tendency for change to be more common where manual unions were recognised. The contrast was especially marked at places employing from 200 to 499 manual workers but it was also apparent where from 50 to 199 manual workers were employed. In cases where 500 or more manual workers were employed there were not enough establishments which did not recognise manual unions for viable statistical comparison, although it remained the case that those few were less likely to have introduced change of any of the three kinds than the very large majority of workplaces of that size where unions were recognised.

We repeated the analysis shown in the first half of Table II.7 separately for manufacturing industry alone and the results are summarised in the second half of the table. It remained the case that workplaces in manufacturing where unions were recognised were more likely to have introduced change, though when the comparison was confined to places of similar size the contrast was less marked than it was across all sectors. It was not possible to take the formal comparison beyond workplaces employing 200 manual workers because so few manufacturing plants of that size did not recognise unions. But the pattern for the 17 plants of that size which did not recognise manual unions was similar to that among smaller workplaces. In the private service sector there was more diversity than in manufacturing in the pattern of union recognition but technical change,

Table II.7 The introduction of advanced technical and other major changes in relation to the recognition of manual unions

Percentages

	All sectors		Manufacturing industry	
	Manual union recognised	Manual union not recognised	Manual union recognised	Manual union not recognised
All establishments				
Advanced technical change	17	11	37	24
Any major change	40	33	70	59
Unweighted base	*1405*	*448*	*426*	*118*
Establishments with 1-49 workers				
Advanced technical change	11	9	29	20
Any major change	31	30	63	58
Unweighted base	*457*	*322*	*56*	*61*
Establishments with 50-199 manual workers				
Advanced technical change	24	20	34	(33)[a]
Any major change	56	40	70	(61)
Unweighted base	*422*	*92*	*149*	*40*
Establishments with 200-499 manual workers				
Advanced technical change	44	(33)	§[b]	§
Any major change	70	(54)		
Unweighted base	*274*	*27*		

[a] See note B.
[b] Unweighted base too low for analyses.

especially change involving microelectronics, was generally less relevant. The pattern of contrasts between places that recognised manual unions and those that did not was, however, generally very similar to that in manufacturing.

Our analysis of the experience of workplaces which did not recognise manual unions compared with those which did showed that there was no sign that the freedom from constraints of trade union recognition enabled managements more readily to introduce new technology. There remained the possibility, however, that in workplaces where unions were represented, the level of union organisation and activity might be associated with the propensity of the organisation to change. Accordingly, we analysed the frequency with which different forms of change were introduced in relation to union density or the proportion of manual workers who were union members. Table II.8 summarises the overall picture. It is apparent that in the economy as a whole there was no consistent association between trade union density and the extent to which workplaces introduced change. Ignoring workplaces where there were no union members, it does look as though there may have been a tendency for our smallest workplaces, in terms of the number of manual workers employed, to have been less likely to have introduced change the higher their level of manual union density. The indications within the next size band, however, were that any tendency was in the opposite direction.

When we focussed our analysis upon manufacturing industry the number of workplaces in different categories in different size bands was often too few for satisfactory comparisons to be made. But there was again the hint, in the one size band with a sufficient range of variation in union density, that microelectronic innovation was, if anything, more common the higher the level of union density. In manufacturing workplaces employing 200 or more manual workers, however, manual union densities were generally so high that it was possible to compare only places with 100 per cent union membership, or what was effectively a closed shop, with workplaces where the large majority but not all manual workers were trade unionists. This limited and unsatisfactory comparison did suggest that there was a slight tendency for change to be less common in places with full trade union membership. Those few larger workplaces with less than 50 per cent manual membership had, however, experienced substantially less change.

In summary then, it appeared that there was no consistent association between manual trade union density and the extent to which workplaces introduced different forms of change but that there was a positive association between manual trade union recognition and the propensity of workplaces to introduce change. The overall pattern certainly provided no evidence that the general effect of manual trade unions was to inhibit managements from introducing new technology or organisational

Table II.8 The introduction of advanced technical and other major changes in relation to trade union density among manual workers

	Manual trade union density					
	No members	1-24	25-49	50-89	90-99	100 per cent
All establishments						
Advanced technical change	9	18	15	19	21	17
Any major change	30	47	37	44	38	43
Unweighted base	*355*	*103*	*96*	*307*	*162*	*597*
Establishments with 1-49 manual workers						
Advanced technical change	7	20	14	14	(11)[a]	11
Any major change	29	42	32	36	(24)	35
Unweighted base	*275*	*51*	*51*	*98*	*35*	*193*
Establishments with 50-199 manual workers						
Advanced technical change	24	(10)	(19)	27	26	26
Any major change	36	(57)	(54)	58	52	57
Unweighted base	*63*	*40*	*29*	*104*	*51*	*158*
Establishments with 200-499 manual workers						
Advanced technical change	§[b]	§	§	38	(60)	46
Any major change				72	(76)	72
Unweighted base				*58*	*38*	*115*
Establishments with 500 or more manual workers						
Advanced technical change	§	§	§	(53)	(60)	66
Any major change				(82)	(92)	82
Unweighted base				*47*	*38*	*131*

[a] See note B.
[b] Unweighted bases too low for analysis.

change. On the other hand, the fact that trade union recognition was positively associated with the introduction of change did not necessarily mean that trade union recognition actively encouraged management innovation. There appear to be three main possible explanations of that association. The first is simply that trade union recognition was co-incidentally associated with the introduction of change as a consequence of some third influence which encouraged both. We showed that the association was independent of size and sector, two of the strongest influences upon both the rate of technological change and levels of trade union recognition, but there may have been others. Secondly, it may well be that in places where unions were recognised there were greater incentives for managers to innovate. That could arise either from pressure from union representatives to change or from a greater attraction to substituting capital for labour. Thirdly, of course, it is possible that in some places the introduction of change was associated with an increase in trade union representation.

The extent of microelectronic applications affecting the jobs of manual workers

Having located advanced technical changes within a framework of the range of major changes affecting manual workers, we now go on to consider the extent to which workplaces used microelectronic applications affecting manual workers. In introducing that analysis we have to emphasise that our survey was not designed to measure the extent to which workplaces used different types of technology. If it had been, then we would almost certainly have selected different respondents on the management side. We would have looked for people with more technical knowledge than the personnel managers who were our most common respondents in the larger workplaces. As it was, our principal focus of interest was industrial relations arrangements and institutions. Accordingly, we selected the management respondent most likely to know about these and associated topics. Our interest in technological change and the extent to which different forms of technology were adopted, chiefly concerned patterns of variation. In particular, we wanted to establish the extent to which the pattern of variation was associated with industrial relations arrangements. We were not aiming for exact measures of how many workplaces in Britain used microelectronic technology or how many employees worked on such technology. Moreover, our interest in the use of advanced technology principally focussed upon process applications or its use in new methods of production or working. Previous research showed that such applications had more important implications for the workers affected and their reactions than the use of microelectronics in new products(2).

We emphasise these points as a preface to our findings on the extent of applications of advanced technology in order to stress that these results should be taken as indications rather than precise measures of the extent to which microelectronic technology has had an impact upon the jobs of workers. We focus a little later on our results from manufacturing industry, where we also interviewed works managers if our primary management respondent was a personnel manager. There we show that the reports of personnel managers appeared to be sounder when they were telling us about process applications than when they were talking about product applications, perhaps reflecting the greater implications of process applications for personnel. Certainly manufacturing industry was the sector of most interest so far as microelectronic applications affecting manual jobs are concerned. Both manual workers and microelectronic applications affecting their jobs tended to be concentrated in that sector. As Table II.9 shows, 43 per cent of workplaces in manufacturing were using advanced technology affecting manual workers compared with 10 per cent in private services and seven per cent in public

Table II.9 Extent of microelectronic applications affecting manual workers in relation to sector

Column percentages

	Total	Private manu-facturing	Private services	Nation-alised industries	Public services
Microelectronic applications affecting the jobs of manual workers	19	43	10[a]	30	7
No applications of advanced technology	80	56	88	68	91
Not stated	1	1	1	2	1
Year of initial application	1980	1978	1981	1979	1981
Base: all establishments employing manual workers					
Unweighted	*1853*	*580*	*515*	*191*	*567*
Weighted	*1749*	*412*	*689*	*103*	*545*

[a] See note C.

services. Just under one third of workplaces belonging to nationalised industries had introduced manual applications. Not only were microelectronic applications more widespread in manufacturing but they had also been introduced at an earlier stage in that sector. Manufacturing workplaces that were using microprocessors had, on average, introduced them in 1978. In services it tended to be three years later before those establishments with advanced technology first adopted it. As the implications of advanced technology for manual workers were so heavily concentrated in manufacturing we shall devote the bulk of our analysis of the distribution of the new technology upon that sector. But, first, it is worth saying a further word about the extent to which manual workers generally were directly affected by microelectronics.

Across the economy as a whole about one fifth of workplaces employing manual workers operated some plant or equipment involving manual workers that included microelectronics (see Table II.10). As applications

Table II.10 Quantitative impact of advanced technology upon manual workers

Percentages

	All sectors	Manufacturing industry
Proportion of workplaces using microelectronics affecting manual workers	19	43
Proportion of manual workers working at places using microelectronics	37	67
Proportion of manual workers themselves working on processes including microelectronics	11	16
Base: establishments employing manual workers		
Unweighted	*1853*	*580*
Weighted	*1749*	*412*

were very much more common in larger workplaces the proportion of blue collar employees who worked at places which used microelectronic equipment rose to over one third. But as only a proportion of the total manual workforce at each location worked on equipment that incorporated microelectronics, the proportion of the total manual workforce personally working on the new technology fell to 11 per cent. In manufacturing the proportions were very much higher. Nearly one third of workplaces had advanced technical processes that affected manual workers. Two thirds of manual workers were employed at places using such equipment. Sixteen per cent of manual workers in manufacturing personally worked on such equipment. The proportion of manual workers engaged in working upon equipment using microprocessors was really very substantial. This is especially true when it is compared with the coverage of, say, mass production in the 1950s(3).

The distribution of microelectronics in manufacturing industry

As we mentioned earlier, manufacturing industry as a whole provided scope for the use of advanced technical components in two main ways. The first was the scope for incorporating microelectronics in products, though that scope was limited to some sectors of manufacturing. The second was the scope for adopting new processes including machines, equipment and plant which incorporated microelectronics. As we also mentioned, previous research has shown that process applications have had more impact upon the workers affected(4). Product applications might involve no more than substituting one microelectronic-based component for an earlier conventional one and have very few implications for the way in which products were made or for the jobs of the people making them. In contrast, as we show in later chapters, process applications often involved changes in working methods, in the content of the jobs of people working on the process, in levels of earnings and in the number of people engaged in the activity involved.

Reference to the previous work on microelectronic applications in manufacturing raises the question of how the pattern of our findings compared with it. By far the most substantial previous research on the use of the new technology in manufacturing industry was carried out by Northcott and his colleagues at PSI(5). In the survey that he carried out shortly before our own, Northcott was concerned exclusively with manufacturing and his sample (1,200) was larger than our sub-sample in manufacturing (561). Moreover, Northcott's survey was specifically designed to measure the extent of microelectronic applications. Hence, he selected the respondents likely to have the most technical knowledge at his workplaces while our respondents were selected because they were likely to be the management representatives most knowledgeable about personnel and industrial relations matters. Consequently, in relation to sample size and the nature of the respondent selected, his results on the extent of applications should have been more reliable than ours with regard to the measurement of the extent of applications. On the other hand, his results were based upon telephone interviews which must be regarded as marginally less reliable than the personal interviews which were the source of our information.

Table II.11 shows the overall picture on manufacturing applications as revealed by the two surveys. It is clear that our findings on process applications look a little low relative to those of Northcott, while our results concerning product applications look a little high. Our survey was carried out one year later than Northcott's, when, according to the trend he identified, process applications would almost certainly have become more common, and our findings do not show such an increase. In contrast, our results show a very substantial increase in product applications and an increase which is almost certainly greater than that which occurred. A second check on our results was provided by the fact that in manufacturing establishments where our primary management respondent was a personnel manager, we also had, in many instances, parallel interviews with the works manager who was asked the same questions on

Table II.11 Microelectronic applications in manufacturing industry: WIRS 1984 compared with Northcott 1983[a]

Column percentages

	WIRS 1984	Northcott 1983
Any application	46	47
Process application	*43*	*43*
Product application	*23*	*10*
No application	54	53
Unweighted	*561*	*1200*
Weighted	*370*	*1890*

[a] The WIRS 1984 subsample was of manufacturing establishments employing 25 people or more. Northcott's sample was of manufacturing establishments employing 20 or more. In each instance the percentages were weighted to be representative of all manufacturing establishments included in the respective samples.

process and product applications. Table II.12 shows the results of the comparison. The first striking feature of the table is that it appears to reveal a much greater use of microelectronics than the general pattern shown in Table II.3. In fact, workplaces where we spoke to both personnel and works managers tended to be larger manufacturing establishments and the use of advanced technology affecting manual workers tended to be concentrated in such workplaces. Secondly, on the assumption that works managers were likely to be better informed than person-

Table II.12 A comparison of works manager and personnel manager accounts of microelectronic applications in manufacturing industry

Column percentages

	Works managers	Personnel managers
Any application	66	60
Process application	*63*	*57*
Product application	*22*	*31*
No application	34	40
Base: manufacturing establishments where both personnel and works managers were interviewed		
Unweighted	*276*	*276*
Weighted	*76*	*76*

nel managers about microelectronic applications, the results of the comparison in Table II.12 were consistent with our interpretation of the contrast between the WIRS 84 results and those of Northcott. They suggest that our primary management respondents were inclined to overstate the extent of product applications but slightly to understate the extent of process applications. It may be, as we suggested earlier, that this was partly a consequence of the greater interest of the personnel function in process applications and also of the order in which we asked the questions. We asked about applications in products first. It may be that a few managers selected principally for their knowledge of personnel and industrial relations matters and less well informed about technical matters, especially insofar as there were differences between products and processes, were inclined to miss the qualifying phrase *in any of your products* at the end of the question.

Whatever the reason for the apparent tendency among our primary management respondents to overstate the extent of product applications and understate the extent of process applications, two points need to be re-emphasised about that tendency. First, as mentioned earlier, the more significant applications of microelectronics for industrial relations purposes are process applications. The answers regarding process applications from our primary respondents appeared more reliable and it was those that we used principally for analysis. Secondly, as we explained at the beginning of this chapter, we were mainly interested in the pattern of variation in the use of advanced technology. Our results appeared perfectly robust for that purpose. At every point where Northcott found significant variations in the extent to which microprocessors were used, that tendency was reflected in our findings. Equally, any source of variation in the distribution of microelectronics revealed by analysis of the answers of our primary management respondents in manufacturing was mirrored in analysis of the answers of works managers.

Variations in applications of microelectronics in manufacturing

Before moving on to consideration of any associations between industrial relations arrangements and the use of advanced technology we need to report the three dominant sources of variation in their use in manufacturing industry. These were size, ownership and industrial sector. First, Table II.13 shows that the larger the number of manual workers employed on site, the more likely were establishments to apply the new technology. This was true for both products and processes, though the trend was most marked in relation to processes. The fact that advanced technology was more common in larger workplaces may appear wholly predictable, since any event or characteristic is more likely to be observed on a purely chance or random basis, the larger the population or area

Table II.13 Microelectronic applications in manufacturing industry in relation to number of manual workers employed

Percentages

		Number of manual workers employed at establishment				
	Total	10-49	50-99	100-199	200-499	500-more
Any application	**46**	**37**	**46**	**59**	**70**	**89**
Process application	43	33	44	55	68	85
Per cent of workforce covered	22	19	25	19	26	24
Year of introduction	1981	1981	1981	1979	1980	1978
Product application	23	24	19	20	25	38
Per cent of workforce covered	35	35	41	25	40	32
Year of introduction	1981	1982	1980	1978	1979	1978
No application	**54**	**63**	**54**	**41**	**30**	**11**
Base: all manufacturing establishments						
Unweighted	*561*	*91*	*83*	*93*	*136*	*158*
Weighted	*370*	*201*	*87*	*43*	*27*	*12*

observed. The table also shows, however, that in workplaces that had introduced microelectronic applications, the coverage of the workforce was just as extensive in proportionate terms as in smaller workplaces. Consequently, manual workers in larger workplaces were substantially more likely to be operating microprocessor-based equipment than counterparts in smaller workplaces. For instance, 89 per cent of establishments employing 500 or more manual workers used advanced technical processes and managers in such establishments estimated that, on average, about one quarter of manual employees worked on such processes. Accordingly, about one fifth of manual workers in the largest establishments worked on plant, machines or equipment that incorporated the new technology. In contrast, only about six per cent of manual workers in small workplaces were engaged in such processes, for only one third of such workplaces had any process applications, and those that did covered on average only 19 per cent of the workforce.

Secondly, and most strikingly, the use of advanced technology was much more common in manufacturing workplaces belonging to overseas companies than in UK counterparts (see Table II.14). The contrast was especially striking so far as small and medium-sized workplaces were concerned. The difference was slightly more marked as regards product applications, but it was also marked for process applications. As in rela-

Table II.14 Extent of microelectronic applications affecting manual workers in manufacturing industry, in relation to nationality of ownership and whether independent or part of group.

Percentages

	UK owned	Foreign owned	Inde-pendent	Part of group
Any application	42	71	35	52
Process application	40	54	33	46
Unweighted base	*426*	*105*	*75*	*441*
Small establishments (25-99 employees)				
Any application	§ [a]	§	(31) [b]	40
Process application			30	35
Unweighted base			*48*	*66*
Medium size establishments (100-499 employees)				
Any application	56	(79)	§	§
Process application	53	71		
Unweighted base	*160*	*36*		

[a] Unweighted base too low for analysis. In order to illustrate the way in which the contrast between UK-owned and foreign-owned establishments and between independent establishments and those that were parts of groups persisted across size bands we have isolated size bands where there were sufficient numbers in each of the categories to make the comparisons.
[b] See note B.

tion to size, our findings on the unfavourable position of domestic operators relative to overseas companies mirrored that identified by Northcott. Moreover, we found throughout our analysis that, according to all our measures of the use of advanced technology and of the pace of different forms of change, overseas companies operating in Britain were strikingly more innovative than domestic counterparts. This was an important finding both in its own right and in relation to its implications for possible explanations of the differential use of advanced techniques, as we discuss in our conclusions. A further feature of ownership in relation to the use of the newest technology was whether it was independent or belonged to a group (see also Table II.14). Manufacturing workplaces that were part of a group were substantially more likely to deploy the new technology than independent outfits, and this remained true regardless of the size of the workplace.

Thirdly, in relation to industrial sector, the new technology was as might be expected most common in engineering and process industries and least common in textiles, clothing and footwear.

Variations in the use of the new technology in relation to industrial relations arrangements

Our findings on the relationship between the use of advanced technology affecting manual workers and industrial relations arrangements and institutions were consistent with our analysis in relation to the introduction of different forms of change reported earlier in the chapter. They may be summed up briefly. We certainly found no sign that levels or forms of trade union organisation at manufacturing workplaces inhibited the use of advanced technology in either processes or products. The relevant aspects of unionisation included whether or not manual trade unions were recognised; trade union density; the existence of a closed shop; the number of manual unions recognised, and the number of bargaining groups for manual workers.

Overall, there was a marked tendency for workplaces that recognised trade unions to be more likely to use advanced technology than those which did not (see Table II.15). This was true for both process and product applications. Moreover, where establishments were using advanced

Table II.15 Microelectronic applications in manufacturing industry in relation to whether or not manual unions were recognised

Percentages

	Manual union recognised	No manual union recognised
All manufacturing establishments		
Any application	52	38
Process application	49	33
Unweighted	*459*	*102*
Weighted	*213*	*157*
Small establishments (25-99 employees)		
Any application	40	35
Process application	38	30
Unweighted	*64*	*59*
Weighted	*114*	*134*
Medium size establishments (100-499 employees)[a]		
Any application	61	(54)[b]
Process application	58	(51)
Unweighted	*166*	*36*
Weighted	*78*	*22*

[a] There were insufficient large workplaces, employing 500 or more people that did not recognise manual trade unions in manufacturing to continue the analysis beyond that in the size bands shown.

[b] See note B.

processes the coverage was as extensive in unionised workshops as in those where no manual union was recognised. This association was predictable, however, in view of the strong and consistent associations between both trade union recognition and the number of people employed and the use of advanced technology and the size of workplaces. But even when we compared plants of similar size it remained the case that those which did not recognise unions were less likely to have adopted the new technology. This analysis was possible only for establishments employing fewer than 500 employees in total, as beyond that number there were insufficient workplaces not recognising unions to provide a viable basis for analysis.

Table II.16 shows the association between manual trade union density (that is to say, the proportion of all manual workers who were trade union members) and the extent to which the new technology was being used. It is clear again that there was no consistent tendency in manufacturing industry for the introduction of microelectronic-based processes or products to be less common the higher the level of trade union membership. There were, however, two notable features in the analysis. There did appear to be a small number of workplaces in the medium sized range, particularly, which had no union members at all but a relatively high level of microelectronic application. In contrast, there was another group where there were some union members but a low density, and the use of new technology was relatively very low indeed. These two contrasting types of workplace, both towards the bottom of the union density scale, certainly demonstrated the lack of any simple association between union density and the adoption of microelectronics. At the other end of the union density continuum were those workplaces with 100 per cent membership among their manual workers. This raised the question of whether having a closed shop which required all workers to be union members represented a qualitatively different arrangement which had repercussions that were quite distinct from those of other points on the density continuum. Certainly, there is no suggestion in Table II.16 that workplaces with 100 per cent membership were less likely than average to have adopted new processes or products incorporating microelectronics. Detailed analysis of workplaces with and without requirements for workers to be trade union members revealed no hint that such arrangements were associated with the propensity of workplaces to use the new technology. Equally, we found no independent associations between the operation of advanced equipment and either the number of manual trade unions recognised or the number of manual bargaining groups.

Table II.16 Microelectronic applications in manufacturing industry in relation to trade union density (manual trade union members as a percentage of manual workforce)

	Manual union density[a]					
	No members	1-24 per cent	25-49 per cent	50-89 per cent	90-99 per cent	100 per cent
All manufacturing establishments						
Any application	36	(23)[b]	(49)	62	41	51
Process application	32	(23)	(49)	56	41	49
Unweighted	*66*	*26*	*26*	*91*	*66*	*242*
Weighted	*119*	*25*	*26*	*60*	*30*	*93*

Small establishments (25-99 employees)

	Manual union density		
	No members or fewer than 25 per cent	25-89 per cent	90-100 per cent
Any application	31	(52)	(35)
Process application	26	(51)	(33)
Unweighted	*52*	*27*	*37*
Weighted	*119*	*52*	*69*

Medium size establishments (100-499 employees)

	Manual union density			
	No members	1-49 per cent	50-99 per cent	100 per cent
Any application	(71)	(30)	62	63
Process application	(66)	(30)	57	62
Unweighted	*20*	*23*	*62*	*81*
Weighted	*15*	*13*	*35*	*32*

[a] Within each size band, the grouping of the densities was selected to reflect the distribution of densities within that size band and the general pattern of microelectronic applications over the range.
[b] See note B.

In view of the scale of our inquiry, the analysis in this section of the extent to which the use of new technology affecting manual workers was associated with trade union organisation has been somewhat limited by the fact that we confined the analysis to manufacturing industry. This limitation did not apply to the early part of the chapter, however, when we were able to look at a range of different forms of change affecting manual workers across all sectors. The pattern of association between the introduction of change and trade union recognition that we identified in that context was largely reflected in the subsequent more narrow analysis of the use of microelectronics in manufacturing.

Footnotes

(1) BBC Radio 4 *Today* programme, 6 March 1986.
(2) Jim Northcott with Petra Rogers, *Microelectronics in Industry: What's Happening in Britain*, PSI, No.603, 1982.
(3) Robert Blauner, *Alienation and Freedom*, University of Chicago Press, Chicago, 1964.
(4) Northcott with Rogers, 1982, *op.cit.*
(5) Jim Northcott and Petra Rogers, *Microelecronics in British Industry: The Pattern of Change*, PSI, No.625, 1984.

III Change and the Use of Computers and Word Processors in the Office

We were impressed, as we reported in the previous chapter, by the extent to which major change was an integral part of working conditions for manual employees. We subsequently found that office workers had become even more subject to major change. Office change was common in all sectors of employment and not concentrated in just one sector. It affected more workers. It was dominated by change involving the new technology. The computer has been a familiar feature of life in business, commerce and industry since the mid-1960s. The development of microelectronics in the mid-1970s, however, gave rise to a new generation of mini and micro computers, smaller, cheaper, more flexible and more powerful than their predecessors. At the same time, microelectronics led to the development and rapid spread of word processors. For our analysis of the introduction of advanced technology in relation to non-manual workplace industrial relations, we focussed upon the introduction of word processors and computing facilities. In order to standardise the category with which we were concerned in different workplaces, when asking about their implications and impact we also focussed upon *office workers*, whom we defined as *secretarial, administrative, clerical and typing staff*. This focus enabled us to analyse the reactions of a broad group of non-manual workers who were common to nearly all the establishments included in our survey whether they were in manufacturing, private services or public services. It meant that we limited our analysis of the impact of new technologies on non-manual workers to one specified sub-group of that broad category, whereas we had looked at new technology in relation to manual workers as a whole. But we judged that some focus was necessary at the non-manual level and what we lost in the comprehensiveness of the coverage we more than gained in the scope that the focus provided for comparison between sectors.

In this chapter, we look at the distribution of computers and word processors and at the introduction of change involving them in much the same way as we did in Chapter II for manual workers in relation to the introduction of advanced technology on the shop floor. One way of structuring our report would have been to take the results concerning manual workers as our starting point and simply to highlight the features of our findings concerning office workers that differed from that model. That approach, however, would have done less than justice to our enquiries devoted to non-manual grades. Traditionally, research and writing about workers and technical change have tended to concentrate on manual workers(1). This was partly because manual workers were affected much more by technological change but also because manual workers and their representatives were felt to be a bigger obstacle to change. Our present findings, however, show the growing importance of technical change in the office. Accordingly, our results concerning office workers warrant presentation and analysis in their own right rather than simply as an adjunct to the experience of manual workers. This may involve some repetition where the pattern of results and variations was common to both levels of employee, but this is unavoidable if we are to do justice to both sets of results. Accordingly, the first part of this chapter looks at the introduction of change involving computers or word processors in the office compared with other forms of change. The second part looks at patterns of variation in the distribution of word processors and computing facilities.

The extent of different forms of change in the office

When asking about our three forms of major change in the office, we substituted *the introduction of word processing or computer applications* for *the introduction of new plant, machinery or equipment that included microelectronics technology* in the equivalent question about change on the shop floor. Hence, we asked managers and workers' representatives whether each of the following three forms of change affecting office workers had been introduced into their workplaces in the previous three years:

i) word processing or computer applications;
ii) new machinery or equipment, not involving word processor or computer applications (excluding routine replacement);
iii) substantial changes in work organisation or working practices not involving new machinery or equipment.

Our results, which are summarised in Table III.1, show that about one half of all workplaces employing non-manual workers had experienced such major change. Change was more common the larger the number of non-manual workers employed. In consequence, nearly three quarters of non-manual workers were employed at places that had experienced major change. The most recent innovation introduced had, on average, directly affected 39 per cent of the office workers at the workplace. As change was more common in the private sector, the figures for that sector are shown separately.

Table III.1 The extent of major change affecting office workers in the previous three years

Percentages

	Any major change involving new equipment or organisation	
	All sectors	Private sector
Proportion of workplaces experiencing change	49	53
Proportion of non-manual workers employed at workplaces experiencing change	71	73
Proportion of office workers at establishment covered by most recent specific change	39	36
	Change involving word processors or computers	
	All sectors	Private sector
Proportion of workplaces experiencing change	35	41
Proportion of non-manual workers employed at workplaces experiencing change	61	66
Proportion of office workers at establishment covered by most recent specific change	36	36

The extent of change experienced by office workers, as revealed by Table III.1, may be compared with the comparable pattern for manual workers shown in Table II.1. It was apparent that on every count office workers experienced more change than manual workers, and the contrast was especially marked so far as advanced technical change was concerned, with office workers about twice as likely to experience such change. Indeed, major change affecting office workers was dominated by the introduction of word processors or computers; for instance, two-thirds of non-manual workers in the private sector were employed at establishments which had experienced such change in the previous three years. Change involving advanced technology was more than twice as common as either of the other two forms of change, as we show later. Moreover, it was common for places to experience more than one type of change (see Table III.2). About one third of non-manual workers were employed at places which experienced at least two or more of the three forms of change; 8 per cent worked where they had experienced all three. In some instances, of course, offices will also have introduced more than one application of a particular form of change.

As stated earlier our detailed analysis of variations in the distribution of computers and word processors comes in the second half of this chapter. Predictably, the types of office that recently introduced advanced technical change tended to be very similar to the types that we later identify as

Table III.2 Number of different major changes in the office

Column percentages[a]

	All establishments	Number of non-manual workers employed 1-24	25-49	50-99	100-199	200-499	500 or more	All non-manual employees
Any one of the three forms of change	35	28	44	38	41	39	37	39
Any two of the changes	12	6	11	18	26	24	37	24
All three forms of change	3	*	2	5	9	14	11	8
None of the three forms of change	51	65	42	39	24	23	15	29
Not stated	1	1	1	–	–	*	–	*
Base: all establishments employing non-manual workers								
Unweighted	*2010*	*463*	*320*	*317*	*311*	*304*	*295*	
Weighted	*1985*	*974*	*507*	*270*	*135*	*69*	*30*	

[a] See note C.

Table III.3 Forms of major office change in relation to number of non-manual employees

Column percentages[a]

	All establishments	Number of non-manual workers employed						
		1-24	25-49	50-99	100-199	200-499	500 or more	All non-manual employees
Change involving word processors or computers	35	20	40	51	67	68	82	61
Change involving conventional new equipment	15	8	18	18	27	32	23	23
Organisational change	16	13	16	20	26	30	39	27
None of the three forms of change	51	65	42	39	24	23	15	29
Not stated	1	1	1	–	–	–	–	*
Base: establishments employing non-manual workers								
Unweighted	*2010*	*463*	*320*	*317*	*311*	*304*	*295*	
Weighted	*1985*	*974*	*507*	*270*	*135*	*69*	*30*	

[a] See note E.

using computers and word processors. The analysis here provides a perspective for that picture by showing the frequencies with which organisational change and conventional technical change were introduced relative to advanced technical change. As already indicated, there was a very strong and consistent tendency for advanced technical change to be more common the larger the number of non-manual workers employed. There was also a consistent but less strong tendency for organisational change to be more common in larger workplaces, but the trend in relation to the introduction of conventional equipment was less clear (see Table III.3). The private sector and manufacturing industry, in particular, were much more active than the public sector in introducing word processors and computers into offices but the public sector introduced conventional technical change and organisational change with a similar order of frequency to that in the private sector (see Table III.4).

Table III.4 Forms of major office change in relation to sector

Column percentages[a]

	All establishments	Private manufacturing	Private services	Nationalised industries	Public services
Change involving word processors or computers	35	47	38	30	23
Change involving conventional new equipment	15	11	15	13	17
Organisational change	16	17	14	20	19
None of the three forms of change	51	43	49	56	57
Not stated	1	*	1	1	1
Base: establishments employing non-manual workers					
Unweighted	*2010*	*592*	*593*	*194*	*631*
Weighted	*1985*	*424*	*836*	*105*	*621*

[a] See note E.

So far as advanced technical change was concerned, differences between sectors became, if anything, more marked when establishments of similar size were compared. Within each size band, the introduction of word processors and computers was most common in private manufacturing industry, less common in private services and least common in public services.

Establishments belonging to groups were substantially less likely to have introduced advanced technical change than the offices of indepen-

Table III.5 Forms of major office change in the private sector in relation to nationality and independence

Column percentages[a]

	UK owned	Overseas owned	Independent establishments	Part of group
Change involving word processors or computers	39	65	41	38
Change involving conventional new equipment	12	26	9	14
Change in organisation only	13	32	9	18
None of the nominated types of change	49	24	49	49
Not stated	*	–	1	1
Base: private sector establishments employing non-manual workers				
Unweighted	*1001*	*184*	*194*	*823*
Weighted	*1147*	*113*	*350*	*764*

[a] See note E.

dent organisations (see Table III.5). They were more likely, however, to be subject to organisational change and to change involving conventional technology. This difference was reflected in a marked tendency for small establishments to be substantially more likely to experience organisational change the larger the size of the organisation of which they were part. But they were less likely to experience advanced technical change. This overall pattern was partly a consequence of the fact that all public sector establishments belonged to larger organisations and there was a distinctive pattern in forms of change in the public sector. Even when we confined our analysis to the private sector, however, it remained true that independent establishments were slightly more likely than those belonging to groups to introduce computers or word processors. The overall contrast was a consequence of the very pronounced difference among smaller establishments. It remained the case, however, that within each size band in the private sector, offices at establishments belonging to larger organisations were substantially more likely than independent counterparts to be subject to organisational change and to introduce conventional technical change.

Establishments belonging to overseas companies were invariably parts of groups. They appeared to derive their propensity to introduce organisational change and conventional technical change partly from that characteristic, and their propensity to introduce advanced technical change from a general tendency, which pervaded our results, for overseas companies to be more progressive than British counterparts. As Table III.5 shows, there were considerable differences between domestic workplaces and those owned by overseas companies. Two-thirds of the latter introduced office change involving computers or word processors, while the comparable figure for British companies was 39 per cent. Moreover, it was not simply that they were more inclined to use the new technology; they were also very much more likely to introduce both organisational change and conventional technical change. The contrasts between domestic and foreign-owned establishments remained very marked when we confined the analysis to private sector workplaces of similar size and excluded independent establishments.

Non-manual trade union organisation and the introduction of change

The chief feature of our analysis of major change affecting office workers in relation to non-manual union organisation was that places where unions were recognised were consistently more likely to experience organisational change than those where they were not. This remained true when establishments of similar size were compared separately in private manufacturing and private services. When we looked at variations in relation to union density, there also appeared to be some indication that organisational change was more common in places with a high level of non-manual trade union membership. Unfortunately, we did not have sufficient numbers in the different cells to sustain the analysis within the different sectors. But there was a hint that the introduction of advanced technical change was less common in those rare instances where there was 100 per cent non-manual union membership. On the other hand, organisational change tended, if anything, to be more common in such workplaces.

It is of particular interest that the introduction of organisational change was more common where trade unions were recognised and where there was a high level of union organisation. In the next chapter we report that all our information on worker and trade union reactions to change combined to show that organisational change, independent of any new equipment or machines, was very much less popular than technical change. And yet we see here that the least popular of all the forms of change included in our analysis was more frequent in places with a relatively high level of non-manual unionisation. It would not be possible to conclude from this that a relatively high level of union organisation was a positive aid to the introduction of less popular forms of change. But cer-

Table III.6 Major changes in relation to non-manual union recognition, sector and size

Percentages

	All establishments		Establishments with 1-49 non-manual workers		Establishments with 50-149 non-manual workers		Establishments with 150 or more non-manual workers	
	Union recognised	Not recognised	Union recognised	Not recognised	Union recognised	Not recognised	Union recognised	Not recognised
All establishments								
Advanced technical change	31	39	18	34	50	60	72	77
Organisational change	19	13	13	10	24	15	34	23
Private sector								
Advanced technical change	47	39	34	34	66	60	83	77
Organisational change	19	13	16	12	24	15	28	23
Private manufacturing								
Advanced technical change	61	42	39	37	78	(81)[a]	89	(82)
Organisational change	21	16	9	16	29	(6)	36	(38)
Private services								
Advanced technical change	41	37	33	32	59	55	74	74
Organisational change	19	12	18	10	21	18	17	16

[a] See note B.

tainly the pattern was not consistent with the view that unionisation represented an obstacle to organisational change.

So far as the introduction of advanced technical change in offices was concerned, there was some initial indication that applications were less common where unions were recognised (see Table III.6). But once again this pattern was largely a consequence of the fact that non-manual unions were better represented in the public sector, and that sector was least inclined to use new technology. When analysis was confined to the private sector, the overall direction of any persistent tendency was for new technology to have been more frequently introduced recently in offices where unions were recognised. With regard to density, there were, as we have said, insufficient numbers to pursue the necessary analysis thoroughly within size bands and sectors. Table III.7 shows the analysis that we were able to do. It reveals the pattern, already discussed, for advanced technical change to be less common where non-manual union membership was very high but for organisational change to be more common.

Table III.7 Major changes affecting private sector office workers in relation to non-manual union density

Percentages

		Density	
	1–49	50–99	100
All private sector establishments			
Advanced technical change	45	48	27
Organisational change	13	16	19
Small private establishments (1–49 manual workers)			
Advanced technical change	36	34	24
Organisational change	11	7	19
Medium size private establishments (50–149 manual workers)			
Advanced technical change	54	69	29
Organisational change	14	28	16
Large private establishments (150 or more manual workers)			
Advanced technical change	83	78	73
Organisational change	19	35	46

The distribution of computers and word processors

We now go on to analyse patterns of variation in the use of modern electronic equipment in offices. We start by identifying the broad framework of variation. Subsequently, we concentrate on exploring the extent to which there has been any association between industrial relations arrangements for non-manual workers and their use of computers and word processors.

Computing facilities at workplaces

Nearly two thirds of all workplaces (62 per cent) employing non-manual workers used one or more computing facilities (see Table III.8). Many used more than one. As might be expected, there was a strong and consistent tendency for this use to increase the larger the size of the establishment as measured by the number of non-manual workers employed. In consequence, the proportion of non-manual workers employed at establishments using computer facilities was 85 per cent, substantially greater than the 62 per cent of establishments with facilities. Having a mainframe computer on site was the type of computing facility having the strongest association with size. In consequence, the contrast between the proportion of workplaces having mainframe computers and the proportion of non-manual workers employed at such workplaces was especially marked.

While the strong tendency for the use of computer facilities to increase with size might have been predictable, three observations may usefully be made about that association. First, the extent to which each and every facility was more common in larger workplaces was striking. It was not the case, overall, that if a workplace had a mainframe computer it was less likely to have micros or to use a bureau service. The opposite was the case. This was a pattern that tended to permeate our results. Those establishments which used one type of computing facility were more likely to use others. It seemed as if the use of one type of facility tended to encourage the use of others. Secondly, the aspect of size with which the presence and number of computing facilities were so strongly associated was the number of non-manual workers employed at the establishment. Table III.8 shows the strength and consistency of the association. In contrast, there was no independent association between the use of computing facilities and the number of *manual* workers at the workplace. Moreover, there was, overall, no independent association between the size of the total organisation of which the establishment was part and the use of computing facilities at a particular establishment. Indeed there was, surprisingly, a slight tendency for computing facilities to be less commonly used at places that were parts of larger organisations. This tendency was more marked in relation to word processing arrangements, and we comment on it further in the next section. Thirdly, the strong positive

Table III.8 Arrangements for computing at establishment in relation to size

Column percentages

	All establishments	Number of non-manual workers employed 1-9	10-24	25-49	50-99	100-199	200-499	500-999	1000 or more	All non-manual employees
Some arrangement for computing at establishment	62	27	55	67	87	91	91	92	98	84
No computing facility	38	73	45	32	12	9	9	8	2	16
Not stated	*	*	–	1	1	*	*	–	–	*
Nature of arrangement[a]										
Mainframe computer in house	9	3	4	6	13	26	31	42	60	27
Link to computer elsewhere in organisation	21	9	11	20	40	47	54	63	63	44
Mini computer in house	23	8	18	27	31	39	50	54	85	44
Micro computer in house	27	5	25	24	45	50	55	63	80	50
Link to computer outside organisation	6	–	2	9	9	13	20	17	30	15
Computer bureau service	10	6	6	10	14	17	23	28	21	18
Base: establishments employing non-manual workers										
Unweighted	*2010*	*165*	*298*	*320*	*317*	*311*	*304*	*175*	*120*	
Weighted	*1985*	*398*	*576*	*507*	*270*	*135*	*69*	*17*	*11*	

[a] See note F.

association between the use of every kind of computing facility and the number of non-manual workers at the establishment hardly suggested that such facilities served as a substitute for non-manual employees. On the contrary, the pattern strongly implied that computing facilities were the tools of non-manual workers. This does not mean that, in particular instances, more staff would not have been needed to carry out particular functions if a computing facility was not performing them, and if it was decided to continue to carry them out. But the strong independent association between the number of non-manual workers employed and all types of computing created the impression that such facilities were favourable to white collar employment, certainly in relative terms. We return to this theme in Chapter IX, when we consider our findings on the implications for employment of different forms of new technology.

Three further general characteristics of workplaces were strongly associated with the extent to which they used computing facilities, namely sector, ownership and industrial category. First, computing facilities were very much less common in the public sector than in private establishments (see Table III.9). In the private sector, they were substantially more widespread in manufacturing than in services. This was true of every type of computing arrangement other than a link to a computer in another part of the organisation of which the establishment was part and, to a lesser extent, a link to a computer in another organisation. The prevalence in financial services and retailing of links across establishments in the same organisation contributed substantially to this pattern. In the public sector, establishments belonging to nationalised industries stood out as places where there were remarkably few arrangements for computing. In contrast, public corporations and quangos used computers in their establishments to an extent that was close to the national average. Public services occupied an intermediate position in the public sector but their use of computing facilities was very low compared with the private sector. The contrasts between the public and private sectors and different parts of the private sector remained marked when differences in size were taken into account; Table III.10 shows the pattern for small establishments. Within each size band, the largest contrast occurred in relation to the existence of a substantial computer, either a mainframe or a mini, on site. There was a marked and consistent decline from manufacturing through private services to public services. The contrast we found between public services and manufacturing industry in their use of microelectronics affecting manual workers was clearly a consequence, at least in part, of the fact that new technology was more relevant to the operation of manaufacturing industry. This was much less true in relation to the use of computing services in the office. Moreover, the general applicability of computing in the office meant that there was no need, as there was with applications of new technology to manual workers, to confine our more detailed analysis of patterns of variation to the manufacturing sector.

Secondly, there was a dramatic difference, in the private sector, between establishments belonging to UK companies and overseas counterparts (see Table III.11). This contrast was apparent within each size band. It tended to be true for every type of computing facility but, again, the difference was especially marked in relation to having a substantial computing capacity, either a mainframe or a mini, at the workplace. The contrast did not result from any tendency for overseas companies to operate in different sectors from British companies or for computing to be more relevant to their operations. In both private manufacturing and private services, workplaces belonging to overseas companies were more

Table III.9 Arrangements for computing at establishment in relation to sector

Column percentages

	All establishments	Private manufacturing	Private services	Nationalised industries	Public services
Some arrangement for computing at establishment	62	75	61	49	55
No computing facility	38	25	39	51	44
Not stated	*	*	1	–	1
Nature of arrangement[a]					
Mainframe computer in-house	9	15	9	4	4
Link to computer elsewhere in organisation	21	19	25	35	15
Mini computer in house	23	35	24	19	15
Micro computer in house	27	32	18	18	36
Link to computer outside organisation	6	5	6	2	7
Computer bureau service	10	14	11	5	5
Base: establishments employing non-manual workers					
Unweighted	*2010*	*592*	*593*	*194*	*631*
Weighted	*1985*	*424*	*836*	*105*	*621*

[a] See note F.

likely to have computing arrangements than establishments of similar size belonging to British companies. And they were especially more likely to have their own on-site computing facility.

Establishments belonging to overseas companies invariably belonged to large groups, which raises the general issue of the rate of innovation at independent establishments as compared with those belonging to groups. We found that the introduction of advanced technology on the shop floor was substantially more likely at workplaces belonging to larger organisations than at independent workplaces. The pattern regarding the use of computers was different; it appeared that independent establishments were more likely to have computing facilities than those belonging to groups, and were particularly more likely to have their own substantial computing facility on site. We consider this difference further when discussing word processing equipment.

Table III.10 Arrangements for computing at small establishments in relation to sector

Column percentages

	Small establishments (1–49 non-manual workers)		
	Private manufac-turing	Private services	Public services
Some arrangement for computing at establishment	69	53	41
No computing facility	31	47	58
Not stated	–	–	1
Nature of arrangement[a]			
Mainframe computer in-house	7	6	1
Link to computer elsewhere in organisation	12	18	6
Mini computer in house	28	22	6
Micro computer in house	26	12	27
Link to computer outside organisation	2	5	4
Computer bureau service	10	8	3
Base: establishments employing non-manual workers			
Unweighted	*220*	*302*	*175*
Weighted	*332*	*654*	*407*

[a] See note F.

Table III.11 Arrangements for computing at establishment in relation to nationality of ownership

Column percentages

	All private sector establishments		Small establishments (1-49 non-manual workers)	
	UK owned	Overseas owned	UK owned	Overseas owned
Some arrangement for computing at establishment	64	83	57	74
No computing facility	36	17	43	26
Not stated	*	–	–	–
Nature of arrangement[a]				
Mainframe computer in-house	9	32	6	17
Link to computer elsewhere in organisation	22	40	14	33
Mini computer in house	27	38	24	24
Micro computer in house	21	45	15	41
Link to computer outside organisation	6	8	4	6
Computer bureau service	12	19	9	13
Base: private sector establishments employing non-manual workers				
Unweighted	*1001*	*184*	*479*	*43*
Weighted	*1147*	*113*	*916*	*70*

[a] See note F.

The third major source of variation in the use of computers lay between the industrial sub-sectors of the broad sectors covered in our earlier analysis. In manufacturing, textiles and leather, footwear and clothing stood out as the industries least likely to use any computing facilities; electrical and instrument engineering and vehicles and transport equipment were those most likely to have their own substantial facility on site. In the private service sector, financial services and business services were markedly more likely than any others to use computers. They were followed by retail distribution and wholesale distribution. Least likely to use computers were hotels and catering establishments and miscellaneous services.

Use of word processors

One quarter of our establishments had word processing equipment, and once again there was a very strong association with the number of non-manual workers employed on site. In consequence, over one half of the non-manual workers at the establishments in our sample worked at places that had word processing equipment. Generally, the pattern of variation in the distribution of word processors closely followed that for computers. The number of non-manual workers employed on site was associated with their distribution in a very strong and independent way, but neither the overall size of the establishment nor of the organisation of which it was part had any independent relationship. Word processors were more common in the private than the public sector and private manufacturing used them more widely than services. In the public sector, nationalised industries operated word processors least frequently, while public corporations and quangos used them relatively widely. One fifth of establishments in the public services had word processors, a proportion which was close to the norm for the public sector. The variations in relation to sector were independent of differences in the characteristic size of establishments in each. Within every size band there was a strong tendency for use to be most widespread in private manufacturing and least common in public services, with private services occupying an intermediate position (see Table III.12). Word processors were much more widely used in foreign-owned establishments than in British counterparts, independently of size and sector. Again the contrast was marked; overall, more than half of overseas-owned workplaces (56 per cent) employed word processors compared with a quarter of domestic counterparts. The concentration of word processors in particular industries and services closely followed the variation in relation to computers.

One feature of our analysis of the distribution of word processors warranted special attention, because it was even more pronounced than the

Table III.12 Proportion of establishments with word processing equipment in relation to size and sector

	Private manufacturing	Private services	Public services
Establishments with 1–49 non-manual workers	22	18	8
Establishments with 50–149 non-manual workers	53	45	41
Establishments with 150 or more non-manual workers	86	59	52

Table III.13 Use of word processors in relation to number of non-manual workers at establishment and size of organisation

	Size of organisation		
	25–499 employees	500–9999 employees	10000 or more employees
Number of non-manual workers employed at establishment			
1–49 non-manual workers	25	14	9
50–149 non-manual workers	51	48	38
150 or more non-manual workers	87	73	54

comparable pattern in the use of computers. As the first part of Table III.13 shows, when establishments employing similar numbers of non-manual workers were compared, establishments were less likely to use word processors the larger the organisation of which they were part. The pattern shown was partly a consequence of the fact that it included public sector establishments, such as those belonging to nationalised industries, few of which used word processors and all of which were part of very large organisations. But even when we confined our analysis to the private sector, there remained no positive, independent association between organisation size and the use of word processors.

Consistent with the pattern in relation to organisation size, word processors were less common in establishments belonging to groups than in independent establishments. This tendency was independent of broad variations between the size and sector of establishments. Again, the pattern is the opposite of that for applications of advanced technology affecting the jobs of manual workers. In that context we suggested that the extra resources of larger organisations might make the dissemination of new machines and methods more rapid throughout their operations. It is surprising that the same principle did not apply to the application of new technology to office working. It would appear that the divisional and head offices of larger organisations pay more attention to the equipment they use in operations carried out by manual workers at workplaces in their organisations than to the equipment used by office workers. This view is consistent with our findings in Chapter V where we show that decisions to introduce advanced technology on the shop floor were more likely to be centralised than those to introduce microelectronic equipment in the office.

Associations between the distribution of computers and word processors and levels of trade union organisation

In principle, we had more scope for analysis of white collar applications of advanced technology in relation to union organisation than for blue collar applications. The possibility of the adoption of word processing equipment or computer facilities was relevant to office working and office workers in all the sectors we covered. In addition, there was more diversity in the private sector with regard to the recognition of unions for non-manual workers in different sectors and in levels of trade union density among non-manual workers. Accordingly, there was more scope for analysing variations on the basis of larger numbers of cases enabling us to place greater confidence in the results. In practice, it soon became clear that the workplace characteristics that dominated the distribution of both computer facilities and word processing equipment were also strongly related to levels of non-manual union organisation. These characteristics included sector, size as indicated by the number of non-manual workers employed, and variations between different industries and services within broad sectors. In consequence, it remained difficult to identify comparable cases, where there were variations in union organisation, through which we could measure any independent influence of union organisation.

Table III.14 shows the details in relation to union recognition. Overall there was very little difference between the proportion of all establishments recognising non-manual unions that used computer facilities (61 per cent) and the proportion of all establishments not recognising unions which did so (63 per cent). When, however, we looked at establishments of similar size, it did appear that places where non-manual unions were recognised were less likely to use computers. But that analysis included public sector offices, which were very much more likely than private sector workplaces both to recognise unions and to work without computers. The pattern in the public sector largely determined the overall picture regarding the use of computers and non-manual trade union recognition. Non-manual unions were so widely recognised in the public sector that it was not possible to compare workplaces where unions were recognised with those that were not. In the absence of such evidence, it would take a very courageous commentator to argue that, among all the characteristics distinguishing public sector offices from private sector counterparts (such as aims, functions, criteria for judging performance, the nature of managers, forms of organisation, levels of decision-making and investment funding), the one characteristic that accounted for the comparatively low usage of computing in public sector offices was the fact that non-manual unions were almost universally recognised there.

This argument became even more implausible in view of our analysis of the private sector as shown in the lower rows of Table III.14. It was

Table III.14 Computing arrangements in relation to recognition of non-manual unions

Percentages

	All establishments		Establishments with 1–49 non-manual workers		Establishments with 50–149 non-manual workers		Establishments with 150 or more non-manual workers	
	Union recognised	Not recognised	Union recognised	Not recognised	Union recognised	Not recognised	Union recognised	Not recognised
All establishments								
Some computing arrangement	61	63	48	57	86	93	91	98
Microcomputer	*31*	*22*	*20*	*19*	*50*	*37*	*61*	*40*
Bureau service	*7*	*12*	*5*	*9*	*9*	*26*	*19*	*32*
Private sector establishments								
Some computing arrangement	73	63	63	57	90	93	100	98
Microcomputer	*25*	*22*	*11*	*19*	*46*	*37*	*65*	*40*
Bureau service	*12*	*12*	*8*	*9*	*15*	*26*	*33*	*32*
Private manufacturing								
Some computing arrangement	81[a]	73	64	70	97	99	100	99
Microcomputer	*12*	*3*	*16*	*28*	*52*	*(46)*[a]	*73*	*(43)*
Bureau service	*18*	*13*	*7*	*11*	*23*	*(30)*	*34*	*(38)*
Private services								
Some computing arrangement	70	58	62	49	86	92	100	97
Microcomputer	*19*	*18*	*9*	*13*	*42*	*35*	*55*	*39*
Bureau service	*10*	*12*	*8*	*9*	*9*	*26*	*32*	*29*

[a] See note B.

apparent that when offices of similar size were compared, those that recognised non-manual unions were more likely to use computers and this pattern remained largely constant when private manufacturing industry and private services were analysed separately. The one exception was provided by the contrast between small service establishments and small manufacturing plants. The general patterns would suggest, however, that the sources of those differences were more likely to be the distinctive features of particular sub-sectors within manufacturing and services, relating to both the use of computers and trade union recognition rather than to the independent influence of non-manual union recognition.

Association between different types of computer facility and union recognition
We have focussed so far on the possibility that there might be associations between the extent to which workplaces used any computing facilities and whether they recognised non-manual unions or not. We found that there was no evidence of any consistent, independent influence. There remained the possibility that there were variations in the type of facility likely to be used by places that recognised unions compared with those that did not. For instance, it appeared overall that places recognising non-manual unions were more likely to have micro computers than those which did not, but less likely to use a bureau service. Insofar as there is a strong tendency for workplace union representatives to favour the carrying out of functions in house rather than contracting them to agents outside, it seemed possible that this might be a difference that was consistently and independently associated with trade union recognition. It soon became apparent, however, as we compared patterns within different sector and size bands, that there was no such association. Similarly, it looked, overall, as if places where unions were not recognised were less likely than those where they were to have a substantial computing capacity on site, either a mainframe or a mini computer. Again, however, this pattern was a consequence of the position in the public sector, where the recognition of non-manual unions was common and having a substantial in-house computer facility rare. When we looked at associations within sectors and size bands we found no consistent pattern. Within one size band, any difference was in one direction; in others it was in another. It was clear that there was no consistent relationship between the type of computing facility used by workplaces and their position in relation to trade union recognition, just as there was no association with whether there were any computing arrangements.

Union density and use of computers
The extent to which workplaces used computing facilities in different sectors and size bands, in relation to the proportion of non-manual work-

Table III.15 Computing arrangements in relation to non-manual union density

Percentages

	Non-manual union members as a proportion of the non-manual workforce				
	1–24 per cent	25–49 per cent	50–89 per cent	90–99 per cent	100 per cent
All establishments					
Some computing arrangement	68	76	67	63	39
Microcomputer	*23*	*34*	*38*	*32*	*18*
Bureau service	*11*	*14*	*7*	*4*	*6*
Private sector establishments					
Some computing arrangement	69	91	74	75	45
Microcomputer	*21*	*38*	*32*	*(14)*[a]	*10*
Bureau service	*13*	*18*	*10*	*(11)*	*9*
Private manufacturing					
Some computing arrangement	82	92	74	89	76
Microcomputer	*27*	*28*	*41*	*(24)*	*(29)*
Bureau service	*9*	*17*	*16*	*(27)*	*(18)*
Private services					
Some computing arrangement	56	90	74	§[b]	42
Microcomputer	*(14)*	*(38)*	*25*		*8*
Bureau service	*(17)*	*(19)*	*5*		*8*

[a] See note B.
[b] Unweighted base too low for percentages.

ers who were trade union members, is shown in Tables III.15 and III.16. Once again it is clear that, when analysis was confined to places that had some non-manual union members, there was no consistent association between union density and use of computers. If anything, the most general pattern appeared to be that computing facilities were most common in places with intermediate levels of non-manual union membership; they were less common in places with a low level of membership or none. It was not possible to pursue our analysis within size bands separately for manufacturing and service establishments, because of the small numbers in many of the categories; but the indications were that there was some persistence to that pattern. Again, given the lack of any consistent associ-

Table III.16 Computing arrangements in relation to non-manual union density and size

Percentages

	Establishments with 1–49 manual workers Density			Establishments with 50–149 manual workers Density			Establishments with 150 or more manual workers Density		
	1–49%	50–99%	100%	1–49%	50–99%	100%	1–49%	50–99%	100%
All establishments									
Some computing arrangements	67	50	33	83	87	76	97	89	93
Microcomputer	24	24	15	32	53	46	62	59	35
Bureau service	10	4	4	17	13	24	32	18	20
Private sector establishments									
Some computing arrangements	76	61	(40)[a]	83	97	(66)	100	100	§[b]
Microcomputer	23	12	(6)	30	60	(28)	67	55	
Bureau service	12	4	(7)	19	11	(23)	35	36	

[a] See note B.
[b] Unweighted base too low for percentages.

ation between union density and the use of computers, the chances are that the tendencies for places with low levels of non-manual union membership to make relatively little use of computers and for those with full non-manual union membership to make similarly little use, were a consequence of other aspects of their circumstances or characteristics.

Word processing facilities and non-manual union arrangements

We also examined the possibility that the use of word processors might be associated with the recognition of non-manual unions or with non-manual union densities. That is to say, we repeated the analysis shown in Tables III.14 to III.16 but focussing upon the use of word processors. Again we found that there was no clear or consistent association, when we took differences in size and sector into account, between non-manual trade union organisation and the extent to which word processors were used. In summary, then, we have identified major differences in the extent to which offices used computers and word processors but the level of trade union organisation at workplaces was not one of them. The strongest and most important sources of variation identified were sector and nationality of ownership. We take up the implications of this pattern of results in our conclusions.

Footnotes

(1) It was not until the first generation of mainframe computers began to spread that technical change in the office became a topic for research; see Enid Mumford and Olive Banks, *The Computer and the Clerk,* Routledge and Kegan Paul, 1967. For a review of developments by the early 1980s, see John Steffens, *The Electronic Office: Progress and Problems,* PSI, 1983.

IV Technical Change Compared with Organisational Change

The most striking and rewarding results of research are often produced by accident. Thus it was with our present results showing the differential impact of different forms of change. As we explained in Chapter I, we decided to make the introduction of change one of the main new substantive areas to be covered in the present survey in our series, and we chose change involving the new microelectronic technology as the form of change upon which to focus. This choice was made not because we expected that such advanced technical change was intrinsically different from any other form of change, but because it was the most topical form of major change and would provide a specific focus for our questions. Privately, we thought, first, that it was only one form of change among many. Secondly, we expected, from our previous research on the management of change and our reading of other people's research, that the reception for any change would depend principally upon its content, to a lesser extent upon the manner of its introduction, but very little upon its form or packaging(1). By the content of the change, we meant its implication for earnings, employment and intrinsic job satisfaction. By the manner of its introduction, we meant the extent to which the people affected by the change and their representatives were involved in decisions about different aspects of the change. By the packaging of the change, we meant whether it took the form of *advanced technical change, conventional technical change or organisational change*, as defined in Chapter I. Indeed, we thought it worthwhile asking about those other forms of change in order to demonstrate that the introduction of microelectronic technology represented just one type of change among many, which was managed in much the same manner as any other and whose reception was dependent upon similar considerations to any other.

We certainly established, as we showed in Chapter II, that although the introduction of advanced technology affecting manual workers was very common in the early 1980s, it was slightly less common than the introduction of conventional technology or of major organisational change. We also showed, in Chapter III, that although the introduction of word processors and computers was the most common form of change in offices over the same period, innovations involving conventional technology or major organisational change were also common. On the other hand, as we report in this chapter, we discovered, contrary to our expectations, that the form of the change did matter. There were inherent differences in the impact and implications of the three different forms of change and especially as between the two types of technical change, on the one hand, and organisational change independent of new machines or equipment, on the other. Indeed, contrasts between the implications of the three distinct forms of change permeated our analysis. In Chapter V, we show that the chief source of variation in the extent to which personnel managers were involved in the management of change was the distinction between whether it was an advanced technical change, a conventional technical change or an organisational change. In the same chapter, we also find that the form of change was a major source of variation in the extent to which shop stewards consulted full-time officials about change. Chapter VII reveals that the introduction of advanced technical change was associated with the same kind of relaxation of traditional demarcations as between different categories of manual worker that was sought through the classical productivity agreements that made up a substantial part of the organisational changes. But the relaxation of demarcations appeared to be very much more acceptable when it was associated with the introduction of new machines than when it was not. In the present chapter we show that there were marked differences between the answers to all our detailed questions about particular changes in relation to the form of the change; that is to say, differences in the extent of support for the change; in assessments of its implications for work content; in levels of consultation and negotiation over the change; in its impact upon earnings and upon manning; and in the level of the organisation at which it was initiated.

The chapter has three purposes. First, it highlights one of the major independent sources of variation in the impact and implications of major change established by our study. Secondly, it prepares the ground for the concentration upon the introduction of advanced technology in subsequent chapters. The major focus of our questions about particular changes was the introduction of new plant, machinery or equipment incorporating microelectronics. In Chapters V to X we provide a detailed examination of different aspects of the introduction of such advanced technology, including the influences upon its introduction; its impact upon job content, employment and earnings; the levels of worker involvement in decisions over change; and the levels of worker and trade union support for advanced technical change. This chapter clears the way for that focus by showing how that particular form of change compared with conventional technical change and organisational change in relation to the issues dealt with in more detail in subsequent chapters. Thirdly, the chapter serves as a general introduction to our findings upon the impact and implications of change before we subsequently analyse those findings in more detail and depth.

Before moving on to those findings, however, we should make clear one of the limitations of the comparisons that we shall be making. The design of our survey meant that we did not collect detailed information about all the organisational changes or about all the conventional technical changes introduced by the workplaces covered in the survey. We asked whether they had introduced each of the three forms of change. If they had introduced new technology based upon microelectronics then we focussed our detailed questioning about the change upon the most recent change taking that form(2). But some establishments that had introduced advanced technology had also introduced conventional technology and organisational change as well. We did not have time in our interview to repeat our questions on the detail of the change in relation to the second two types of innovation. Similarly, if establishments had not recently introduced advanced technology but had acquired new conventional technology, we asked detailed questions about the most recent change that had involved new technology and ignored any organisational change which they reported. It was only if establishments had introduced no new technology, either advanced or conventional, but had experienced organisational change, that we asked about the details of the organisational change. Consequently, the cases of organisational change about which we collected full details represented only a sub-sample of all the cases of organisational change reported in the survey. In fact, as we showed in the previous chapter, recent organisational change was reported at 19 per cent of establishments, but only 10 per cent of establishments experienced only organisational change of the three forms of innovation we identified. In consequence, we collected the details of about only one half (54 per cent) of the cases of organisational change identified in the survey. We cannot know how far that sub-sample was representative of the total population of organisational changes because they occurred in distinctive establishments under distinctive circumstances. Similarly, we collected details of about only two thirds (68 per cent) of the conventional technical changes reported in the survey because the other one third had also introduced microelectronic technology. These were necessary limitations on our analysis imposed by the time available for interviews.

Despite any limitations in our data about the implications of different forms of change, our analysis of those implications proved very convincing. The contrasts between the different forms of change as revealed by our analysis were both substantial and robust. At each stage we checked to ensure that the differences persisted when we compared establishments of similar size, in similar sectors, having similar forms of union organisation and subject to similar trends in the size of their workforces.

Our most striking finding about the different forms of change was that there was very much more support, from both the workers affected and from their trade union representatives, for both forms of technical change than for organisational change. This was true according to the accounts of personnel managers, general managers, works managers and shop stewards. It applied similarly to changes affecting manual workers and changes in the office. It persisted independently of the implications of the change for manning and employment. Technical change in a context of declining employment was more acceptable than organisational change in circumstances of employment growth. We start the analysis of differences between the three forms of change by focussing on changes affecting manual workers and showing the much greater popularity of technical change over organisational change and then go on to explore differences in relation to the other questions we asked about change. We end the chapter with a necessarily more brief consideration of the implications of different forms of change in the office.

Differential levels of support for different forms of change affecting manual workers

In Chapter VIII we analyse in detail the extent of worker and trade union support for advanced technical change both in the office and on the shop floor. Here, we focus simply on the extent of support for such change compared with levels of enthusiasm for conventional technical change and the very much more equivocal reactions to organisational changes. We focus initially upon different forms of change affecting manual workers. As we show later in the chapter under a separate heading, the pattern regarding changes in the office was not dissimilar, but change in the office was so dominated by the introduction of advanced technical change relative to the other forms that there was less scope for satisfactory analysis in relation to the different forms.

Table IV.1 shows the reactions of both workers and shop stewards to the three different forms of change, according to managers' accounts. The chief feature of these reports, as we have already indicated, was how much more hostile both workers and stewards were to organisational change than to change that involved either form of technical change. We consider in Chapter X some of the complexities of identifying reactions to change. Here we are simply concerned with the contrasts apparent between the different forms. It is clear that they were gross. In the majority of cases, managers attributed resistance to workers and shop stewards in their responses to organisational change. In a substantial minority of cases, the resistance was described as strong. By contrast, managers reported in the large majority of cases that workers had supported changes involving new technology. In only a very small proportion of cases was strong resistance reported. It is also apparent, however, that according to

Table IV.1 Reactions of manual workers and shop stewards to different forms of change, according to managers

Column percentages[a]

	Total	Manual workers affected			Manual shop stewards		
		Advanced technical change	Conventional technical change	Organisational change	Advanced technical change	Conventional technical change	Organisational change
Strongly in favour	37	36	56	10	37	44	15
Slightly in favour	32	36	26	31	35	29	26
Slightly resistant	21	18	14	36	25	23	39
Strongly resistant	5	2	1	17	2	3	21
Not stated	6	8	2	6	–[b]	–	–
Support score[c]	+80	+90	+120	20	+60	+80	20
Base: establishments with 25 or more manual workers reporting major change							
Unweighted	*909*	*478*	*258*	*173*	*478*	*258*	*173*
Weighted	*647*	*255*	*215*	*178*	*255*	*215*	*178*

[a] See note C.
[b] In many cases, shop stewards were judged by managers not to have been involved and where the reactions of stewards were not given they are excluded from the table.
[c] See note I for details of the calculation of this score.

managers' accounts, the workers directly affected were substantially more strongly in favour of conventional technical change than of advanced technical change.

It might appear from Table IV.1 that, while shop stewards were equally as doubtful as workers about organisational change, they were slightly less enthusiastic than their constituents about technical change. This pattern, however, was derived from managers' reports. When we looked at what stewards themselves had to say about their reactions and those of workers, we found that their separate accounts of their own feelings and those of the workers directly affected by change, suggested that stewards were more strongly in favour. We discuss in our later analysis of support for change possible reasons for dissonance between the perceptions of managers and those of shop stewards regarding the feelings of workers towards change compared with those of stewards. In the context of this chapter, the main point about stewards' reports is that they revealed a very similar pattern of reaction to the different forms of change to that apparent from the reports of managers (see Table IV.2). A refinement that we were able to build into our interviews with stewards was to distinguish between initial reactions to news of the proposed change compared with subsequent feelings at the time of interview after changes had generally been implemented. This refinement enabled us to identify that, according to stewards, initial reactions to the prospect of organisational change tended on balance to be slightly unfavourable. In contrast, both they and their constituents were generally in favour of both forms of technical change from the start. It was clear from the average scores at the foot of the table, that there was increased support for each of the three forms of change over the period between the two points in time. But it was also apparent that the contrast between initial reactions and subsequent feelings was especially strong in relation to advanced technical change. Thirty-eight per cent of the workers affected were felt to be doubtful about the introduction of advanced technology when it was first proposed, but that proportion fell to a very small seven per cent at the time of interview. Experience of advanced technical change seemed to dispel doubts more effectively than with the other forms of change. Of course, so far as conventional technical change was concerned, there were many fewer initial doubts to dispel.

It is plausible that the more widespread initial doubts about advanced technical change arose from lack of familiarity with the new machinery which led to concern about the implications of the unknown. It appeared that with familiarity came reassurance. This possibility raises the general question of why both types of technical change were so much more popular than organisational change. Some of the reasons will become apparent from our analysis later in the chapter. We show, for instance, that organisational changes were more likely than technical changes to have

Table IV.2 Manual shop stewards' accounts of their reactions and those of manual workers to different forms of change

Column percentages[a]

	Total	Manual workers affected			Manual shop stewards		
		Advanced technical change	Conventional technical change	Organisational change	Advanced technical change	Conventional technical change	Organisational change
Initial reactions							
Strongly in favour	25	22	40	14	34	48	30
Slightly in favour	13	18	18	2	19	18	8
Mixed feelings	22	22	17	27	21	15	15
Slightly doubtful	17	22	16	10	17	13	7
Very doubtful	23	16	9	45	10	5	41
Not stated	–	–	–	–	*	*	*
Support scale[b]	*	+10	+80	–100	+60	+110	–20
Subsequent feelings							
Strongly in favour	37	32	53	27	46	58	33
Slightly in favour	28	31	29	17	29	26	14
Mixed feelings	24	29	14	28	15	12	30
Slightly doubtful	10	4	2	23	7	2	20
Very doubtful	2	3	2	2	3	1	2
Not stated	1	1	1	3	*	1	2
Support scale	+110	+120	+150	+60	+130	+160	+80
Base: manual shop stewards who reported major change							
Unweighted	*549*	*288*	*135*	*126*	*288*	*135*	*126*
Weighted	*267*	*101*	*80*	*86*	*101*	*80*	*86*

[a] See note C.

involved both reductions in manning and in earnings. Those, of course, were important findings in their own right, but they also began to remove the mystery about why so many organisational changes should have provoked doubts or resistance. They did not, however, remove all the mystery. The tendencies for establishments experiencing organisational change to be more likely than those introducing technical change to have suffered both specific reductions in their workforces and enforced redundancy over the previous year were slight (see Table IV.3). Moreover, we

Table IV.3 Extent of manpower reductions and redundancy at establishment in relation to different forms of change

Column percentages

	All establishments	Advanced technical change	Conventional technical change	Organisational change
Whether any manpower reduction				
Yes	41	41	43	55
No	59	59	57	45
Not stated	*	*	*	–
Whether redundancy				
Enforced redundancy	10	13	9	19
Voluntary redundancy	8	17	12	14
Base: all establishments employing manual workers				
Unweighted	*1853*	*478*	*258*	*173*
Weighted	*1749*	*255*	*215*	*178*

carried out our analyses of reactions to different forms of change separately for workplaces whose workforces had been declining and for those that had been stable or growing. We found that the contrast between the unpopularity of organisational change and the popularity of technical change still persisted. For instance, we summarise in Table IV.4 managers' accounts of the reactions of the workers directly affected to the three different forms of change when they were being introduced, in relation to different employment circumstances. First, we show reactions in places where employers introduced particular workforce reductions in the previous year, compared with reactions where there were no such reductions. Secondly, we show reactions in places where the overall size of the workforce had fallen over the previous four years compared with

Table IV.4 Managers' accounts of workers' reactions to different forms of change, in relation to movements in the size of the workforce

Scores and percentages

	Advanced technical change	Conventional technical change	Organisational change
Whether specific manpower reduction at establishment in previous year			
Workforce reduction (support score)[a]	+70	+110	−60
Proportion slightly or strongly resistant (per cent)	*29*	*22*	*66*
No workforce reduction (support score)	+110	+130	+20
Proportion slightly or strongly resistant (per cent)	*15*	*12*	*40*
General movement in size of workforce over previous four years			
Workforce smaller (support score)	+80	+120	−50
Proportion slightly or strongly resistant	*29*	*20*	*62*
Workforce stable or larger (support score)	+100	+110	*[b]
Proportion slightly or strongly resistant	*14*	*17*	*48*

[a] See note I.
[b] Neutral.

reactions in circumstances where the workforce increased or remained stable.

It is apparent that each form of change was more acceptable when it was introduced in a favourable climate of employment so far as movements in the establishment's workforce were concerned. The difference was especially marked in relation to specific reductions in the workforce during the previous year. At the same time it is equally apparent that the general order of popularity of the three different forms of change remained very similar whatever the context so far as employment was concerned. Advanced technical change appeared to provoke slightly more resistance than conventional technical innovation. But change in organisation or working practices independently of technological innovation provoked very substantially more resistance than either form of technical change. Indeed, organisational change appeared to result in substantially more resistance when introduced under favourable employment conditions than technical change introduced in unfavourable employment circumstances. The differences were marked and the implications profound.

The pattern of response to different forms of change in relation to varying circumstances regarding movements in the size of the workforce certainly disposed of any suggestion that the unpopularity of organisational changes was simply a consequence of the fact that they tended more frequently to lead to job loss. The fact that technical change associated with job loss was substantially more popular than organisational change associated with growth or stability in employment showed that technical change did have inherent attractions independent of its implications for employment. We repeated our analysis of reactions to the three forms of change excluding cases where they led to reductions in pay and reductions in manning. We found that technical changes remained substantially more popular. Indeed, the pattern of popularity that we identified in relation to the three forms of change proved remarkably robust throughout our analysis, independently of whether we were analysing the reports of managers or shop stewards; or whether we were looking at the reactions of workers or shop stewards; or whether we were focussing upon office change or innovation on the shop floor; or whether we were looking at reactions in workplaces of different size and sector. It did appear that technical change was intrinsically more attractive than organisational change.

The impact of different forms of change upon pay and manning levels

Few issues have generated more heat than the implications of advanced technology for employment. We report in some detail the contribution that our study has to make to that debate in Chapter IX. In Chapter X we go on to analyse the more neglected issues concerned with the implications of advanced technology for the earnings of the workers affected. Here, we focus upon two narrow issues. We examine, first, how far specific advanced technical changes were more or less likely to result in reductions in manning in the sections of manual workers affected, compared with conventional technical change and organisational change. Secondly, we look at the immediate implications of the three different forms of change for the pay of the manual workers affected. The results are summarised in Table IV.5. Perhaps the most striking initial feature is the extent to which all forms of innovation led to no change in either pay or manning in most cases. Secondly, where there was a change in pay it was substantially more likely to be an increase than a decrease. This was especially true for both forms of technical change. As mentioned in the previous section, organisational change was much more likely than the

Table IV.5 Implications of different forms of change for pay and manning levels, according to managers

Column percentages[a]

	All changes	Advanced technical change	Conventional technical change	Organisational change
Implications of change for pay of manual workers in section(s) affected				
Increased	24	24	24	25
Decreased	4	2	*	14
No change	71	73	76	60
Not stated	*	1	*	1
Implications for manning levels				
Increased	14	11	19	12
Decreased	25	19	19	46
No change	61	70	61	42
Not stated	*	*	*	*
Base: establishments with 25 or more manual workers reporting one or more changes				
Unweighted	*817*	*458*	*224*	*135*
Weighted	*468*	*212*	*152*	*103*

[a] See note C.

other forms of change to result in a reduction in earnings for the manual workers affected. Nevertheless, according to the accounts of managers, it remained more common for organisational change to result in increases than in loss of earnings.

Thirdly, the three different forms of change did have substantially different implications for manning. Conventional technical change appeared to lead to increases in manning in the section affected as often as it led to decreases. Advanced technical change led to reductions more often than increases. And, as highlighted in the previous chapter, direct job loss most frequently resulted from organisational change. Indeed managers reported reductions of manning in nearly one half of the cases of organisational change. The figures in Table IV.5 are based upon the reports of managers and, as we discuss in Chapter X, there was some disagreement between managers and stewards in their separate accounts of the implications of advanced technical change for the earnings of manual workers. Nevertheless, the broad pattern of contrasts revealed by the accounts of manual shop stewards of the impact of different forms of change upon earnings and manning broadly reflected the differences shown in Table IV.5.

The implications of different forms of change for job content

All three forms of change tended on balance to lead to increased skill, responsibility, task repertoire and intrinsic job interest for manual workers, according to the reports of both managers and shop stewards. Table IV.6 shows the answers of managers. Although the impact of changes tended on balance to be favourable, it was not invariably so. For instance, managers reported that advanced technical change had a favourable impact on most aspects of jobs more frequently than any other form of change. On the other hand, in other cases, it also had an unfavourable impact more frequently upon all dimensions, other than the level of supervision. In consequence, when both favourable and unfavourable effects were taken into account, the balance of impact, as measured by our average scores at the foot of Table IV.6, was remarkably similar to that attributed to conventional technical change. Indeed, so far as the implications of the change for workers' control over the pace and content of their jobs were concerned, the balance was to the advantage of conventional technological change. A very slightly different pattern emerged from the independent reports of manual shop stewards of the implications of the three forms of change for job content. According to the accounts of stewards, both forms of technical change had very much more favourable implications than organisational change. So far as the contrast between advanced and conventional technical change was concerned, stewards reported that advanced technical change had a more positive impact upon levels of skill and responsibility and task repertoire, while conventional technical change had more favourable implications for control over the pace and method of working. We discuss more fully in Chapter VI the way in which the different perspectives of managers and shop stewards contributed to slightly different accounts of the implications of change for job content.

Levels of consultation and negotiation over different forms of change

Overall we found that there was surprisingly little negotiation and consultation with manual workers and their representatives over the introduction of advanced technical change in view of the high level of consensus over the desirability of worker involvement in decisions about the introduction of change. There was a tendency for managers to consult only if they were required to do so by the existence of representative institutions or by signs that the change was not immediately acceptable.

Table IV.6 Managers' accounts of favourable and unfavourable changes to specified aspects of manual workers' jobs following innovations

Percentages and scores

	All changes	Advanced technical change	Conventional technical change	Organisational change
Favourable change (percentages)				
More interest	38	46	36	28
Greater skill	34	42	32	21
Wider range of tasks	37	38	36	35
More responsibility	31	33	28	31
More say over how they do their job	25	21	24	34
More control over pace of work	23	22	23	27
Less supervision	17	19	16	13
Unfavourable change (percentages)				
Less interest	7	11	3	6
Less skill	11	15	10	6
Narrower range of tasks	11	15	9	7
Less responsibility	7	12	3	5
Less say over how they do their jobs	20	31	14	7
Less control over pace of work	20	28	15	14
More supervision	19	16	18	26
Average scores[a]				
Interest	+60	+70	+70	+40
Skill	+50	+50	+40	+30
Range of tasks	+50	+50	+50	+60
Responsibility	+50	+40	+50	+50
How they do their jobs	+10	−20	+20	+50
Pace of work	+10	−10	+20	+30
Supervision	*[b]	+10[c]	*	−30
Base: establishments with 25 or more manual workers reporting one or more changes				
Unweighted	*817*	*458*	*224*	*135*
Weighted	*468*	*212*	*152*	*103*

[a] See note I.
[b] Neutral score.
[c] A positive score implies *less* supervision.

We describe and discuss these findings in Chapter VI. They show that consultation and negotiation over the introduction of change affecting manual workers were more common when change was less popular. The pattern of consultation and negotiation in relation to our three forms of change was consistent with that tendency (see Tables IV.7 and IV.8). Organisational changes provoked most doubts, reservations or resistance. They were also subject to the most consultation and negotiation with union representatives. Advanced technical change raised more doubts initially than conventional technical change and formal consultations over advanced technical change were substantially more common. Conventional technological change was most popular and apart from discussions with individuals there was relatively little consultation over such change. Overall, the pattern provided one of the first pieces of evidence to support our conclusion that levels of consultation and negotiation over change affecting manual workers owed more to the extent to which changes were initially acceptable than to the scale of the change and the scope for a useful contribution to the detail of the change from workers or their representatives.

The extent of negotiations over the three different forms of change is shown in Table IV.8. There we report answers to a question which sought to establish how far major change was subject to joint regulation, in the sense that it required the agreement of union representatives, and how far there was simply consultation with union representatives. The proportion of organisational changes subject to joint regulation was three times that of technical changes. This pattern links up with the relatively great extent to which full-time officers were involved in organisational changes. It supports the general impression that we developed that many of the organisational changes were classical productivity agreements(3). This certainly appeared to be the case for many of the organisational changes in manufacturing industry which we discuss in Chapter VII. The pattern also underlines how comparatively free from joint regulation was technical change, even new, unfamiliar and advanced technical change.

Influences on decisions about change from outside the establishment

Decisions to introduce change were frequently taken at a level above that of the establishment, but managers less frequently turned to consultants outside their own organisation for guidance in the introduction of change. We explore influences upon the introduction of advanced technical change in our next chapter. Here again, we simply record that it appeared to be organisational change rather than technical change that was more likely both to be subject to central direction and to lead managers to seek help from consultants.

Table IV.7 Consultations with manual workers and their representatives over the change in relation to different types of innovation

Column percentages

	All experiencing changes				Manual union recognised			
	All changes	Advanced technical change	Conventional technical change	Organisational change	All changes	Advanced technical change	Conventional technical change	Organisational change
One or more of listed forms of consultation	86	82	83	97	91	92	86	96
No consultation	13	17	15	3	8	8	13	4
Not stated	1	1	2	*	1	*	1	*
Formal union channel								
Discussions with stewards	37	39	29	44	52	53	46	59
Discussions with full-time officers	18	16	8	35	25	21	14	50
Formal non-union channel								
Discussions in JCC	15	17	9	21	19	19	13	27
Discussions in ad hoc committees	10	11	8	14	12	13	10	14
Informal								
Informal discussions with individuals	59	58	63	53	54	61	55	39
Meetings with groups of workers	37	38	32	43	40	42	38	36
Base: all establishments with 25 or more manual workers and reporting one or more changes								
Unweighted	*817*	*458*	*224*	*135*	*675*	*389*	*173*	*113*
Weighted	*465*	*212*	*152*	*103*	*317*	*154*	*92*	*71*

Table IV.8 Basis of relationship with manual union representatives over different types of innovation

Column percentages

	All changes	Advanced technical change	Conventional technical change	Organisational change
(a) **All relevant establishments**				
Change negotiated	8	6	4	17
Consultations with union representatives	43	46	39	45
No discussion with union representatives	19	23	19	11
Did not apply/not stated	30	25	38	27
Base: establishments with 25 or more manual workers and reporting one or more changes				
Unweighted	*817*	*458*	*224*	*135*
Weighted	*465*	*212*	*152*	*103*
(b) **Establishments recognising manual unions**				
Change negotiated	11	8	7	24
Consultations with union representatives	61	61	62	60
No discussions with union representatives	27	30	30	16
Not stated	1	1	1	*
Base: as above but confined to establishments recognising manual unions				
Unweighted	*675*	*389*	*173*	*113*
Weighted	*317*	*154*	*92*	*71*

Ninety per cent of our workplaces were parts of large organisations rather than independent enterprises. We reported earlier that establishments that were part of groups were more likely than independent establishments to introduce change affecting manual workers. It is likely that the influence of the head or divisional offices played a part in this tendency, by identifying appropriate new forms of technology and causing them to be diffused among their subsidiaries more quickly than is possible for isolated entrepreneurs or managers at an equivalent level in independent establishments. Certainly, Table IV.9 shows that in the majority (60 per cent) of cases where establishments were part of groups, managers at a higher level than the establishment played a part in the introduction of advanced technical change. In about one half of the cases,

Table IV.9 Levels of management decision-making in relation to different forms of change affecting manual workers

Column percentages[a]

	All changes	Advanced technical change	Conventional technical change	Organisational change
Decision to introduce innovation taken by management:				
— at establishment	42	39	52	36
— at higher level	45	51	32	49
— jointly	12	9	14	15
Other answer/not stated	1	1	2	*
Decision on how to introduce the change:				
— at establishment	57	61	65	41
— at higher level	27	30	13	36
— jointly	12	8	13	18
Other answer/not stated	4	*	9	4
Base: establishments employing 25 or more manual workers having change and part of groups				
Unweighted	*745*	*423*	*194*	*128*
Weighted	*380*	*178*	*110*	*92*

[a] See note C.

higher level management decided upon the innovation. In a further nine per cent of cases, the local manager described the decision as a joint one. Higher level managers clearly played a larger part in decisions to introduce advanced technology than to introduce new conventional plant, machines or equipment. The implication was that group structures did play a part in the diffusion of advanced technical change but when that technology became conventional, decisions might be left more to local managers.

It also appeared, however, that higher level managers played an even more substantial part in the introduction of organisational change than in the introduction of either form of technical change. We have shown that such decisions were relatively unpopular among the manual workers affected. That unpopularity was often reflected also in the feelings of local managers. As we show in Chapter VIII, the reactions to change of first-line managers tended to move closely in line with those of workers. It looked likely that many organisational changes consisted of requirements for savings or greater efficiency imposed upon establishments from the centre. This was further reflected in the way that local managers were not only less frequently responsible for decisions to introduce organisational change but were also very much less likely to have autonomy in its implementation (see the second half of Table IV.9). It appeared that higher level managers were also slightly more likely to play a part in the implementation of advanced technical change compared with their roles in the introduction of conventional technical change.

The pattern of dependence upon higher level managers in their own organisation in relation to the three forms of change was reflected in the extent to which local managers sought help from outside their own enterprises in introducing the change (see Table IV.10). As would be expected, local managers turned to outside consultants more frequently with regard to advanced technical change than to conventional technical change. But they used external consultants most frequently in the introduction of organisational change. Here we should stress that we phrased our question about the use of external consultants in terms of 'how the change should be introduced so far as manual workers were concerned' rather than in relation to any technical advice or assistance. The contrasts in Table IV.10 form part of a general pattern within which substantially more management attention was paid to the human implica-

Table IV.10 Whether anybody outside the establishment was consulted about the change affecting manual workers apart from higher-level office

Column percentages[a]

	All changes	Advanced technical change	Conventional technical change	Organisational change
Yes — consultation with outside body	23	23	15	34
No consultation	76	75	84	65
Not stated	1	2	1	*
Base: establishments with 25 or more manual workers and reporting one or more changes				
Unweighted	*817*	*458*	*224*	*135*
Weighted	*468*	*212*	*152*	*103*

[a] See note C.

tions of organisational change than to those of technical change, and even of advanced technical change. This was also apparent in the comparative levels of involvement of professional personnel managers in the introduction of different forms of change, as we show in Chapter V, and also levels of involvement in the introduction of different forms of change among Head Office managers.

Different forms of change in the office

We mentioned at the start of the chapter that we had less scope to explore the impact and implications of different forms of change in the office than on the shop floor. This was because change in the office over the period with which we were concerned was so dominated by advanced technical change in the form of computerisation and the introduction of word processors. It is true that conventional technical change and organisational change were also common in the office. But, as we also explained in outlining the limitations of our analysis in relation to different forms of change affecting manual workers, we did not ask our detailed questions about particular changes in relation to all forms of change introduced at particular workplaces. If establishments introduced advanced technical change we asked our questions about workers' reactions, consultation and so on, in relation to their most recent change of that type. But if they introduced conventional change we asked detailed questions about that only where they experienced no advanced technical change. Similarly, if they introduced organisational change we asked the full set of questions about that only if they introduced neither form of technical change. Owing to the prevalence of word processing and computing changes, we consequently collected information about only 50 per cent of the conventional technical changes introduced into offices and only 41 per cent of the organisational changes. These proportions meant that our analysis of the differential impact and implications of different forms of change in the office was less satisfactory than the corresponding analysis among manual workers. On the other hand, the analysis that we were able to carry out was of interest and revealed a picture broadly in line with that revealed by the analysis of change affecting manual workers. In consequence, we illustrate that pattern by reporting briefly on differential levels of support among office workers and non-manual shop stewards for the three forms of change. Subsequently, we comment on any notable differences between the office patterns compared with the manual contrasts.

The reactions of office workers and non-manual shop stewards to different forms of change

Office workers and non-manual stewards welcomed both forms of technical change but were more equivocal about organisational change according to the accounts of managers (see Table IV.11). Indeed, the pattern was very similar to that among manual workers and stewards. Conventional technical change was slightly more popular than advanced technical change, but, certainly so far as the office workers affected were concerned, this was largely a consequence of the larger proportion of office workers who were described as strongly rather than slightly in favour of conventional technical change.

It also appears, incidentally, from Table IV.11, first, that office workers were generally more enthusiastic than manual workers about all three forms of change. Secondly, the initial impression created by the table is that non-manual stewards were much less happy about the changes than the office workers directly affected. The contrast between the reactions of workers and stewards certainly appeared much greater than was the case with manual workers (shown in Table IV.1). This impression, however, was based upon the reports of managers. When we looked at the accounts of non-manual stewards of their own reactions and of those of office workers, we found that stewards tended to feel more favourable to each form of change than the office workers affected (see Table IV.12). We analyse in Chapter X the reasons why there should have been so much more dissonance between the independent accounts of managers and stewards in relation to advanced technical change in the office. The second notable feature of Table IV.12 is that organisational change appeared to be even more unpopular relative to technical change, according to the accounts of non-manual stewards, than it appeared from the reports of managers. This was especially so with regard to the initial reactions of stewards and workers to organisational change in the office. The balance of feeling among both stewards and office workers was initially hostile to such change, according to the accounts of stewards.

Other characteristics of different forms of change in the office

The indications were that there was a similarly modest contrast between our other measures of the differential impact of different forms of office change and the corresponding pattern for changes affecting manual workers. First, so far as the implications for staffing and earnings were concerned, organisational change in the office revealed a similar tendency to be more likely to result in job loss than either form of technical change, and for advanced technical change to result in reductions in levels of staffing more frequently than conventional technical change. It was rare, however, for any form of change in the office ever to lead directly to loss of earnings for the office workers affected. In consequence, there was not the same tendency for organisational change in the office to lead more frequently than technical change to loss of earnings. Indeed, organisational change in the office led to increases in earnings in one third of cases, while this was true for one quarter of cases of advanced

Table IV.11 Reactions of office workers and shop stewards to different forms of change in the office, according to managers

Column percentages[a]

	Total	Office workers affected			Non-manual shop stewards		
		Computers or word processors	Conventional technical change	Organisational change	Computers or word processors	Conventional technical change	Organisational change
Strongly in favour	39	38	63	20	25	66	4
Slightly in favour	36	40	21	26	45	10	33
Slightly resistant	15	14	9	32	24	23	40
Strongly resistant	2	2	*	9	5	1	24
Not stated	7	8	6	13	–[b]	–	–
Support score[c]	+100	+110	+150	+20	+60	+120	−50
Base: establishments with 25 or more non-manual workers and reporting one or more changes							
Unweighted	*1123*	*977*	*76*	*70*	*977*	*76*	*70*
Weighted	*639*	*500*	*79*	*60*	*500*	*79*	*60*

[a] See note C.
[b] In many cases, non-manual shop stewards were judged by managers not to have been involved, and where the reactions of stewards were not given they are excluded from the table.
[c] See note I.

forms of change

Column percentages[a]

	Office workers affected			Non-manual shop stewards		
	Computers or word processors	Conventional technical change	Organis-ational change	Computers or word processors	Conventional technical change	Organis-ational change
Initial reactions						
Strongly in favour	22	(9)[b]	(10)	34	(12)	(10)
Slightly in favour	21	(18)	(14)	22	(16)	(57)
Mixed feelings	21	(25)	(23)	13	(22)	(4)
Slightly doubtful	20	(15)	(48)	16	(21)	(16)
Very doubtful	16	(32)	(6)	15	(30)	(6)
Not stated	–	–	–	1	–	(7)
Support score[b]	+20	–60	–30	+50	–50	+60
Subsequent feelings						
Strongly in favour	33	(10)	(21)	44	(14)	(38)
Slightly in favour	26	(24)	(51)	21	(19)	(51)
Mixed feelings	30	(37)	(7)	23	(25)	(9)
Slightly doubtful	4	(18)	(2)	6	(26)	(2)
Very doubtful	6	(11)	(19)	6	(17)	–
Not stated	1	–	–	*	–	–
Support score	+110	+10	+60	+120	–20	+140
Base: non-manual shop stewards who reported major change						
Unweighted	*500*	*26*	*23*	*500*	*26*	*23*
Weighted	*161*	*24*	*20*	*161*	*24*	*20*

[a] See note C.
[b] See note B.
[c] See note I.

technical change but only for five per cent of conventional technical changes. Secondly, it looked as though advanced technical change in the office had slightly more favourable implications for job enrichment relative to conventional technical change, compared with the pattern for changes affecting manual workers. Thirdly, with regard to consultation, there again appeared to be most formal discussions over organisational change and least over conventional technical change. Finally, with regard to the level of centralisation over the initiation of change, there appeared to be an even stronger tendency for decisions to introduce organisational change in the office to be centralised. In contrast, as we discuss in the next chapter, decisions to introduce advanced technical change in the office tended more frequently to be taken locally than was the case for corresponding changes on the shop floor.

Footnotes

(1) W.W Daniel and Neil McIntosh, *The Right to Manage*, Macdonald, London, 1972.

(2) Because of the limits on space in our interviews with principal management respondents, we collected information on the details of the most recent major change (including the level in the organisation at which it was initiated; extent of different forms of consultation with workers and their representatives; extent of negotiations with union representatives; implications for staffing and earnings in section(s) affected; implications for job content; and the reactions of different categories) for only workplaces where 25 per or more manual workers were employed (or non-manual workers in the case of change in the office). Accordingly, we ascertained whether establishments experienced each of the three different forms of change in all cases. We collected information about the details of the most recent major change only where 25 or more of the relevant groups were employed (see structure of interview schedule, page 293, Appendix A for details). In practice that meant that we carried out questioning on the most recent advanced technical change in 70 per cent of cases affecting manual workers and in 60 per cent of cases affecting office workers (weighted figures).

(3) W.W. Daniel and Neil McIntosh, *Incomes Policy and Collective Bargaining at the Workplace*, PEP, No.541, 1973.

V Levels of Decision-making about Technical Change and the Role of Personnel Managers

Our study was focussed upon workplaces or, more technically, establishments. But no workplace is an island. Many were parts of larger organisations. Indeed, even among our private sector workplaces, 72 per cent belonged to companies that operated a number of different establishments. In the public sector, our workplaces tended generally to be parts of large organisations with many operating units. The extent to which local workplaces tend to be part of larger organisations led some commentators to criticise the first survey in this series on the grounds that we focussed on the workplace rather than upon the company or the larger organisation of which the establishment was part(1). Of course, this criticism resembles the comment of the husband who gave his wife two frocks, one blue and one red, for her birthday. When she appeared at dinner proudly wearing the blue frock, he furiously demanded, 'What's wrong with the red one?' Our 1980 survey was unprecedented in its comprehensiveness, and in the sophistication with which it provided a representative national sample of establishments which could be used to make statements about both workplaces and the general population of employees who worked at them. It took into account the perspectives both of different management functions and of worker representatives at the establishment. It was unreasonable to expect such a survey also to be integrated into a representative national survey of the companies and larger organisations of which they were part. We did, however, seek in our present survey, wherever possible, to place local decisions in the context of the constraints imposed by parent bodies. That is to say, when, for instance, we were looking at the introduction of new technology we sought to establish how far this arose from the independent initiatives of local managers and how far it was a consequence of decisions taken by managers at a higher level in the organisation and operating from another establishment. Moreover, all workplaces including independent establishments also operated in an environment of other organisations that might influence them either consciously or indirectly. These included customers, suppliers, the manufacturers of new equipment and the range of organisations in society that seek to bring about change and reform, including government bodies as well as independent consultants and commentators.

The trade unions represented within workplaces were, like the bulk of managements, parts of national organisations. Local trade union representatives operated within the framework of a set of national union policies which might or might not have influence upon the positions that they adopted on any particular local initiative. National union organisations and their regional structures also provided local union negotiatiors with advisory and consultative services that they could call upon when faced with the prospect of new developments at the workplace.

In this chapter we take a limited look at the way influences external to the workplace impinged upon the parties. So far as managers were concerned, we look at the extent to which the decision to introduce change was taken at the local or a more central level in their organisations. We also look at the extent to which they called upon bodies external to the organisation for help or advice in introducing the change. So far as unions were concerned, we look at the extent to which shop stewards consulted the offices of their national organisations and at the responses they received. We start the analysis by examining external influences upon decisions affecting the introduction of advanced technical change affecting manual workers.

The level of decisions to introduce microelectronic technology and the involvement of third parties

Eighty-four per cent of our establishments that introduced microelectronic change were parts of large organisations rather than independent enterprises. We saw earlier that workplaces belonging to groups were substantially more likely than independent workplaces to operate new technology affecting manual workers. This pattern suggested that the greater management and other resources of larger organisations may contribute to the more rapid spread of new methods throughout their operations. An alternative explanation was, of course, that the quality of local managers in terms of their disposition to innovate was higher in places belonging to groups than in independent single establishment firms. Such organisations were almost invariably small commercial organisations in our sample.

Our findings on the level at which the decision was taken to introduce the most advanced technical change certainly suggested that head, divisional or branch offices played a substantial part in major technological

Table V.1 Levels of decision on the introduction of advanced technical change in relation to sector

Column percentages[a]

	Total	Private manu-facturing	Private services	Nationa-lised industries	Public services
Decision to introduce change					
Establishment level	39	59	22	20	(20)[b]
Higher level	51	27	76	79	(79)
Joint decision	9	13	2	1	(1)
Not stated/other	1	1	–	–	–
Decision about how to introduce change					
Establishment level	61	80	41	26	(61)
Higher level	30	11	46	70	(36)
Joint decision	8	7	13	3	(4)
Not stated/other	*	1	–	–	–
Base: establishments with 25 or more manual workers, experiencing advanced technical change and part of group					
Unweighted	*425*	*261*	*51*	*50*	*44*
Weighted	*179*	*87*	*39*	*23*	*24*

[a] See note C.
[b] See note B.

innovation (see Table V.1). The perspectives of our local principal management respondents might have tended to encourage them to emphasise local autonomy(2). Even so, about one half attributed the decision to introduce the technological changes to management at a higher level in their organisations. In 39 per cent of cases it was reported that local managers made the decision and in the remainder the decision was described as a joint one. Greater autonomy was attributed to local managers with regard to decisions about the implementation of the change, as opposed to decisions to make the change. Nevertheless, in nearly one third of cases it was reported that managers at a level higher than the establishment also decided upon how to introduce the change.

There was a clear contrast between private manufacturing industry and all other sectors in relation to the degree of local autonomy. In both private services and all parts of the public sector, decisions on the introduction of change were attributed to higher level managers in three quarters or more of cases. Moreover, in private services and the public sector as a whole, our local managers reported that higher level managers were also responsible for implementing the change more frequently than local managers. There was a particularly high level of centralisation in the nationalised industries. In over two thirds of cases higher level managements decided not only to make the technical change but also how the decision should be implemented.

Differences between the sectors were partly but not wholly explained by size. As might be expected, there was a strong general tendency for establishments to have more local autonomy the larger they were. This tendency was strongest, however, for private and public services. In private manufacturing, local autonomy was relatively high whatever the size band. In services there was a very strong tendency for decisions to become more local the larger the size of the local workforce. In consequence, contrasts between the sectors became less marked in the largest workplaces where new technology had been introduced (see Table V.2).

Table V.2 Decisions on the introduction of advanced technical change at larger workplaces in relation to sector

Column percentages[a]

	Large establishments (employing 175 or more manual workers)			
	Private manufacturing	Private services	Nationalised industries	Public services
Decision to introduce change				
Establishment level	57	(54)[b]	(32)	(53)
Higher level	29	(44)	(63)	(41)
Joint decision	13	(2)	(4)	(6)
Not stated/other	1	–	(1)	–
Decision about how to introduce change				
Establishment level	73	(59)	(54)	(74)
Higher level	10	(39)	(35)	(21)
Joint decision	13	(2)	(9)	(6)
Not stated/other	–	–	–	–
Base: establishments with 175 or more manual workers, experiencing advanced technical change and part of group				
Unweighted	*194*	*20*	*45*	*28*
Weighted	*26*	*4*	*8*	*6*

[a] See note C.
[b] See note B.

Centralisation and trade union representation

As there was a strong association between sector, centralisation of decisions and size, it was difficult to identify any independent association between levels and forms of local union organisation and levels of management decision-making. The indications were, however, that any such independent associations were not strong. There was a hint that, where manual unions were recognised, decisions to introduce change were more likely to be centralised but decisions on the implementation of change were more likely to be local. As Table V.3 shows, in medium-

Table V.3 The extent to which decisions about introducing advanced technology were centralised in relation to trade union recognition

Percentages

	Medium size workplaces	
	Manual union recognised	Manual union not recognised
Decision to introduce change centralised	49	(38)[a]
Decision on implementation centralised	22	(27)
Base: establishments with 50-199 manual workers, experiencing advanced technical change and part of group		
Unweighted	*93*	*20*
Weighted	*59*	*12*

[a] See note B.

sized establishments (employing from 50 to 199 manual workers), the decision to introduce the change was more likely to be centralised in cases where manual unions were recognised than in cases where they were not. But the pattern regarding decisions on implementaion was in the opposite direction. There was some implication here that the existence of local trade union organisation made it less likely that decisions to change would be imposed from the centre. But the pattern in Table V.3 was based upon just one size band, within which the number of establishments not recognising unions was small. The pattern revealed in that size band was not fully reflected in all the others. There remained a tendency, however, in cases where unions were recognised, for there to be a stronger contrast between the level at which decisions to change were taken and the level at which issues about implementation were settled.

Finally in this section, it is clear that decisions to introduce technological innovation that operated to the disadvantage of manual workers were much more likely to be taken at the centre. First, nearly all the changes that resulted in reductions in earnings for the workers involved were decided upon at the centre. Secondly, there was a tendency for changes that resulted in reductions in manning among the workers affected also to be decided at the centre.

The involvement of other third parties

For local managers at workplaces belonging to larger groups, the chief source of outside help, information and advice tends to be specialists at a more central level in their organisation(3). But for purposes of exploring the influences upon managerial innovation it was also of interest to identify what other outside bodies managers relied upon for help or advice about how to introduce change. The most striking difference in the extent to which managers consulted any outside agency about how to introduce change affecting manual workers, and in the sources of their information, lay between managements which recognised unions and those which did not. First, in places where managers recognised manual trade unions, they were more likely to consult outside bodies about the implementation of the change. Secondly, and, on the face of it, more surprisingly, the outsiders whom they most frequently consulted were full-time trade union officers (see Table V.4). They relatively rarely turned to management consultants. Managements which did not recognise unions, however, used management consultants with a frequency similar to that with which counterparts with unions depended upon their full-time union officers. The tendency for full-time trade union officers frequently to act as consultants to management is one that has been identified previously in our series(4). On reflection, it is less surprising that full-time union officers should be used by managers in this way. They have wide experience of new developments in a range of organisations and workplaces, and they are likely to have a good awareness of the types of problem that have arisen and how they have been resolved.

Analysis of sources of advice in relation to manual union density showed that management consultation with full-time union officers was concentrated in establishments where union membership was high. It was also the case that where both full-time officers were consulted by managers and union density was high, the short-term outcomes of the change tended to be more favourable for the workers affected. They were substantially more likely to have received pay increases as a result of the change. The balance of outcomes in relation to manning was also more favourable. Manning was reduced in a similar proportion of cases but it was increased in a higher proportion. But, as we have already indicated,

[illegible] Extent and nature of outside consultations about how to introduce advanced technical change in relation to trade union recognition

Column percentages[a]

		All establishments		Establishments with 50-199 manual workers	
	All establishments	Manual union recognised	Manual union not recognised	Manual union recognised	Manual union not recognised
Outside body consulted	23	25	18	25	(10)[b]
No outside consultations	75	73	82	74	(90)
Not stated	2	3	–	1	–
Nature of body consulted[c]					
Suppliers/manufacturers of plant equipment	7	6	8	3	–
Full-time trade union officers	7	9	–	11	–
Consultants	4	2	9	1	(8)
Management	*3*	*1*	*9*	*1*	*(8)*
Computer	*1*	*1*	–	–	–
Company name given — no indication of function	3	4	–	5	–
Other answer	5	7	–	6	(2)
Not stated	*	*	–	–	–
Base: establishments with 25 or more manual workers and experiencing advanced technical change					
Unweighted	*458*	*405*	*53*	*100*	*26*
Weighted	*212*	*160*	*52*	*64*	*18*

[a] See note C.
[b] See note B.
[c] See note F.

the involvement of full-time officers was only one of a number of inter-related features of the cases that led to favourable outcomes for the workers.

Level of decision-making concerning the introduction of advanced technical change in the office

Decisions over the introduction of advanced technology affecting office workers were substantially more likely to be taken locally than similar decisions affecting manual workers. In over one half of the cases of office

Table V.5 Level of decisions about the introduction of advanced technical change affecting manual workers compared with decisions affecting office workers

Column percentages

	All relevant establishments		Establishments where advanced technical change affected both categories of employee	
	Manual workers	Office workers	Manual workers	Office workers
Decision to introduce the change taken by the management				
-at establishment	39	53	49	61
-at higher level	51	33	37	30
-jointly	9	9	11	7
Other answer/not stated	1	5	3	2
Decision on how to introduce the change taken				
-at establishment	61	64	71	74
-at higher level	30	21	16	12
-jointly	8	11	11	12
Other answer/not stated	*	4	2	2
Base: see footnote [a]				
Unweighted	*425*	*905*	*311*	*311*
Weighted	*179*	*420*	*70*	*70*

[a] The first two columns compare levels of decision-making about all applications affecting manual workers at establishments belonging to larger organisations with the corresponding pattern for all applications affecting office workers. In the second two columns, the comparison is confined to establishments where there were separate advanced technical changes affecting both categories. We adopt a similar format in all our tables comparing experiences and practices relating to manual workers with those relating to office workers. The first contrast provides an opportunity to compare the experiences of office workers and manual workers overall. The second contrast makes it possible to identify the extent to which the same managements dealt differently with manual and office workers.

change, where establishments belonged to groups, managers reported that the decision to introduce word processors or computers affecting the jobs of office workers was taken locally. This proportion compared with 39 per cent in the case of comparable decisions affecting manual workers (see Table V.5). Levels of decentralisation were more similar in relation to the process of implementation. The fact that decisions on the introduction of new technology into offices were relatively decentralised may have made some contribution to the tendency identified in Chapter III for places that were parts of groups to lag behind independents with regard to office applications but to be ahead of independents in applications affecting manual workers. This pattern could owe something to a tendency on the part of managements to feel that changes affecting manual workers were more likely to generate industrial relations problems and, hence, decisions about them could less safely be left to local managers. Alternatively, it might simply be that office applications of new technology were less costly and more frequently came within the budgets of local managers.

Although decisions to introduce new technology in offices were relatively decentralised, it remained the case that many such decisions were taken at higher levels. In one third of cases managers reported that the decision to introduce change was taken at a level more central than the establishment. In a further nine per cent of cases managers reported that the change was the result of a decision taken jointly between local and central managers. As might be expected, there was some tendency for decisions to be more local the larger the number of non-manual workers employed on site (see Table V.6). But this association was not as strong and consistent as the equivalent trend for manual workers. Indeed, it was apparent that the overall organisation size had a stronger influence than the number employed at the establishment upon the level at which decisions about advanced technical change in the office were taken (see Table V.7). The larger was the size of the whole organisation of which the establishment was part, the more likely were decisions to be centralised, independently of the size of establishments. This pattern was not simply a consequence of the fact that the public sector was dominated by large organisations. It remained similar when analysis was confined to the private sector.

So far as sectoral differences were concerned, decisions were most likely to be local in private manufacturing, just as they were in relation to the introduction of change on the shop floor (see Table V.8). Comparing private with public services revealed, surprisingly, that centralised decision-making about office changes was more common in the private sector. The nationalised industries, however, again stood out as organisations where decisions on the introduction of change were most likely to be centralised. In nearly two thirds of the cases of advanced

Table V.6 Decisions on the introduction of word processors and computers affecting office workers

Column percentages

	Number of non-manual workers at establishment						
	All establish-ments	25-49	50-99	100-199	200-499	500-999	1000 or more
Decision to introduce change							
Establishment level	53	48	52	62	47	58	81
Higher level	33	31	41	27	38	33	11
Joint decision	9	11	6	8	11	6	6
Not stated/other	5	10	1	3	4	3	2
Decision about how to introduce change							
Establishment level	64	56	66	73	58	68	88
Higher level	21	22	25	13	25	15	6
Joint decision	11	12	9	11	12	13	3
Not stated/other	4	10	–	3	5	4	3
Base: establishments employing 25 or more non-manual workers that introduced word processors or computers and were part of a group							
Unweighted	*905*	*103*	*150*	*207*	*219*	*135*	*91*
Weighted	*420*	*150*	*119*	*83*	*47*	*13*	*9*

Table V.7 The extent to which decisions to introduce computers or word processors were centralised in relation to establishment and organisation size

Percentages

	Number of non-manual workers at establishment		
	25–49	50–149	150 or more
Total number of employees in organisation			
500–9999	45	43	29
10000 or more	59	47	41

technical change in nationalised industry offices, managers described the decision as having been taken centrally or jointly by central and local managers. The differences in levels of centralisation as between sectors

Table V.8 Decisions on the introduction of word processors or computers in relation to sector

Column percentages[a]

	Private manufacturing	Private services	Nationalised industries	Public services
Decision to introduce change				
Establishment level	65	48	42	52
Higher level	24	38	44	31
Joint decision	8	8	13	9
Not stated/other	3	5	1	7
Decision about how to introduce change				
Establishment level	80	57	59	59
Higher level	6	29	28	20
Joint decision	12	9	8	14
Not stated/other	2	5	5	7
Base: establishments with 25 or more non-manual workers that introduced word processors or computers and were part of a group				
Unweighted	*327*	*213*	*105*	*260*
Weighted	*100*	*180*	*24*	*117*

[a] See note C.

generally persisted when establishments of similar size were compared, with one exception. So far as the smallest workplaces were concerned, those employing fewer than 50 non-manual workers, decisions were very much more centralised in the public services. Above that threshold, and leaving aside nationalised industries, centralisation was most pronounced in private services and local decisions were only slightly more common in private manufacturing than in public services. Differences between domestic and foreign-owned companies were unusually slight – with a tendency for decisions in overseas-owned companies to be more frequently local or joint so far as larger workplaces were concerned. But this pattern was reversed in smaller workplaces.

Levels of decision over office change in relation to non-manual union organisation and the implications of the change for jobs

So far as any association between levels of decision-making and union organisation was concerned, it appeared, in the private sector, that the recognition of unions tended to be associated with a relatively high level of centralisation over the introduction of new technology. This was certainly true so far as smaller workplaces were concerned. It became increasingly less true the larger the number of non-manual workers employed. There appeared also to be an association between the centralisation of decision-making and levels of non-manual density. The associations between centralisation and local trade union organisation were in the same direction as those operating for manual workers.

As with changes affecting manual workers, we looked at any association between the level of the decision over the change and the implications of the change for earnings. We did not find the same pattern of variation. Non-manual applications very rarely resulted in reductions in earnings for office workers and whether the change had a neutral or favourable impact on the pay of the office workers affected did not seem to be associated with the level at which it was initiated. As with the comparable manual changes, however, that minority of cases which did lead to direct reductions in manning among the workers affected were more likely to have been centrally initiated.

Consultations with outside bodies over change in the office

Managers tended more frequently to consult bodies or agencies outside their own establishment or organisation about the introduction of advanced technical change in the office than they did to make such consultations about change on the shop floor (see Table V.9). Establishments of intermediate size were most likely to consult outside bodies, especially those employing 200 to 499 non-manual workers (see Table V.10). Public sector establishments were more likely than private counterparts to consult externally, and they were particularly more likely to consult full-time union officers. Both the propensity to consult and the propensity to call upon full-time union officers were most common among nationalised industries. Private sector managements very rarely called upon full-time officers for help or advice in relation to office applications of new technology; this was in marked contrast to their practice so far as manual workers were concerned. Instead, they were more likely to depend upon the suppliers or manufacturers of equipment for help and advice. The contrast between the public and private sectors was especially strong when larger workplaces were compared. So far as differences between foreign-owned and British companies were concerned, however, and between independents and those that were parts of groups, it was in the middle-sized range that differences were most marked. Offices that were parts of groups were very much more likely than independents to consult outside and overseas companies were substantially more likely than domestic counterparts to do so. Moreover, overseas companies were especially more likely to call upon the services of special-

Table V.9 External consultations over advanced technical change affecting manual workers compared with similar changes affecting office workers

Column percentages

	All relevant establishments		Establishments where advanced technical change affected both categories of employee	
	Manual workers	Office workers	Manual workers	Office workers
Outside body consulted	23	32	22	25
No outside consultations	75	66	75	74
Not stated	2	2	3	1
Body consulted[a]				
Suppliers/manufacturers	7	12	7	10
Management consultants	3	4	1	2
Computer consultants	1	10	*	10
Full-time trade union officers	7	4	9	3
Company name given but no function	3	2	2	1
Other body	5	6	5	5
Base: see footnote [b]				
Unweighted	*458*	*977*	*326*	*326*
Weighted	*212*	*500*	*78*	*78*

[a] See note F.
[b] See note to Table V.5.

ist, independent consultants, whether management consultants or computer consultants. British establishments, in contrast, were more likely to depend upon suppliers or manufacturers.

Analysis of the use of consultants in relation to trade union recognition did not produce the same stark contrasts as the parallel analysis for manual workers. There we found a pattern suggesting that in unionised workplaces, full-time trade union officers fulfilled a similar function to independent management consultants in non-union shops. So far as office workers were concerned, managers were more likely to use consultants if they recognised non-manual unions. This was true of all but the smallest workplaces; there, irritatingly enough, the difference was in the opposite direction, suggesting that the pattern had little direct relationship with union recognition as such.

Table V.10 External consultations about how to introduce technical change in offices in relation to size

Column percentages[a]

		Number of non-manual workers at establishment					
	Total	25-49	50-99	100-199	200-499	500-999	1000 or more
Outside body consulted	32	29	34	34	40	23	10
No outside consultations	66	67	65	64	59	76	90
Not stated	2	4	1	2	2	1	–
Nature of body consulted[b]							
Suppliers/manufacturers of plant equipment	12	13	10	10	15	7	2
Full-time trade union officers	4	2	3	3	13	6	2
Consultants							
Management	4	1	8	5	2	4	–
Computer	10	14	8	8	7	4	2
Company name given – no indication of function	2	1	3	3	1	1	*
Other answer	6	6	6	8	8	5	4
Not stated	*	1	–	–	–	–	–
Base: establishments with 25 or more non-manual workers that introduced word processors or computers							
Unweighted	*977*	*129*	*165*	*220*	*223*	*137*	*103*
Weighted	*500*	*201*	*137*	*91*	*47*	*14*	*10*

[a] See note C.
[b] See note F.

Extent to which shop stewards sought advice and help from outside the establishment

Commentators have tended to identify the growth of *company unionism* as one of the most marked and important trends in the development of British industrial relations over recent years(5). The term is used to identify a growing tendency for trade union members and lay officers within workplaces and organisations to relate their activities, positions, priorities and affiliations to the domestic policies and circumstances of those workplaces rather than to the policies and structure of the wider union of which they are part. Accordingly, it is argued, shop stewards have tended increasingly to operate independently of their national unions and, for instance, to act collectively with representatives of other trade unions within the framework of joint shop stewards committees based upon the employing organisation, rather than with representatives of their own union working at other places within the framework of their own union structure. Previous surveys have found, for example, that the full-time officers of unions have rarely been consulted over periodic plant bargaining about rates of pay(6). Indeed, they tended to show that full-time officers were rarely consulted over anything. In view of these findings and the argument derived from them, we found a surprisingly high tendency for shop stewards to consult full-time officers of their union about the major changes recently introduced at their workplaces. This was especially true so far as manual trade unions were concerned. Manual shop stewards consulted full-time officers regarding about half the major changes that we identified. They turned to officers of their wider union for help or guidance very much more frequently than they consulted any other party. Indeed, full-time officers of their union advised them much more frequently than all other agencies put together. Rather surprisingly, non-manual stewards tended to seek advice or help from their unions less frequently than manual counterparts. They turned to full-time officers in a little over one third of cases. But it remained the case that they also used their own unions for help much more often than they used all other agencies.

Of course, the fact that workplace trade union officers often turned to full-time officers of their unions for help or advice when faced with major change did not necessarily invalidate the general argument about company unionism. We have no information from our present survey about the nature of the guidance that stewards sought from full-time officers. In particular, we could not tell whether they were consulting about what stance to adopt, in the light of general union policy, towards the proposals for change, or whether they were asking for specific information or technical services to support the stance that they adopted unilaterally. If the second practice was more common our findings would still be consistent with the view that trade union decision-making at the workplace has increasingly tended to become independent of wider trade union policies and structures. On the other hand, our present findings suggest that workplace representatives depend upon full-time officers for some kind of help or advice more frequently than has often been suggested by other findings.

Consultation with full-time officers by manual shop stewards

While manual shop stewards frequently consulted full-time officers about the introduction of advanced technical change, they sought help or advice over that form of change less frequently than when faced by the prospect of organisational change (see Table V.11). On the face of it, this was

Table V.11 Extent to which manual shop stewards consulted full-time officers over the introduction of different forms of change

Column percentages[a]

	Total	Advanced technical change	Conventional technical change	Organisational change
Full-time officer consulted	49	41	40	67
No consultations	50	57	58	33
Not stated	1	1	2	-
Level of contact[b]				
Local	26	21	22	36
Regional	19	15	12	30
National	12	11	6	18
Base: manual shop stewards who reported a major change in the previous three years				
Unweighted	*549*	*288*	*135*	*126*
Weighted	*267*	*101*	*80*	*86*

[a] See note C.
[b] See note F.

surprising. It might have been expected that organisational changes, proposed independently of any plan to introduce new plant or machines, would tend to have been mainly composed of familiar types of change with which workplace union representatives frequently dealt; on the other hand, the prospect of advanced technology with mysteriously powerful new microelectronic components, whose implications had been subject to widespread and sensational discussion, might have been expected to lead them to consult full-time officers with experience of the introduction of such technology elsewhere. In fact, manual stewards reported that they consulted full-time officers in two thirds of the cases of organisational change but in fewer than one half of the cases of advanced technical change (41 per cent). Indeed, union officers were contacted no more frequently over the introduction of advanced than of conventional technical change. In view of the picture that emerged in the previous chapter showing mixed feelings about organisational change compared with enthusiasm about technical change, the pattern is less surprising. We reported that organisational change was very much less popular than technical change among both workers and shop stewards. In consequence, the less favourable reactions to organisational change clearly led stewards to seek the help of full-time officers more frequently in dealing with such forms of change. Nevertheless, reactions to the introduction of advanced technology are put into perspective, when we see that they led stewards to consult full-time officers less frequently than they did, for instance, over productivity bargaining changes that have been familiar since the 1960s.

When we focussed specifically upon consultations over the introduction of advanced technical change in relation to sector and size, we found the patterns shown in Table V.12 and Table V.13. It is clear, first, that stewards tended to consult full-time officers much more frequently in the nationalised industries than they did in any other sector. This reflected a generally marked tendency towards centralisation which we have repeatedly identified among the nationalised industries in our study. On the

Table V.12 Extent to which manual shop stewards consulted full-time union officers over the introduction of advanced technical change

Column percentages[a]

	Total	Private manufacturing	Private services	Nationalised industries	Public services
Full-time officer consulted	41	38	(24)[b]	(66)	(32)
No consultations	57	61	(74)	(33)	(65)
Not stated	1	*	(2)	(1)	(3)
Level of contact[c]					
Local	21	25	(14)	(23)	(12)
Regional	15	11	(13)	(22)	(20)
National	11	8	(4)	(27)	(5)
Base: manual shop stewards who reported an advanced technical change					
Unweighted	*288*	*166*	*29*	*49*	*44*
Weighted	*101*	*49*	*16*	*23*	*14*

[a] See note C.
[b] See note B.
[c] See note F.

Table V.13 Extent to which manual shop stewards consulted full-time union officers over the introduction of advanced technical change

Column percentages[a]

	Total	Number of manual workers employed at establishment			
		under 100	100-199	200-499	500 or more
Full-time officer consulted	41	37	(40)[b]	50	55
No consultations	57	63	(58)	48	43
Not stated	1	–	(3)	2	2
Level of contact[c]					
Local	21	14	(23)	35	37
Regional	15	17	(7)	14	22
National	11	11	(10)	10	15
Base: manual shop stewards who reported an advanced technical change					
Unweighted	*288*	*51*	*49*	*74*	*114*
Weighted	*101*	*56*	*20*	*17*	*8*

[a] See note C.
[b] See note B.
[c] See note F.

other hand, the generally marked tendency for workplace representatives to consult full-time officers more frequently than we would have expected from previous findings, was not simply a consequence of the inclusion of the public sector in our study. Our analysis has consistently shown that decisions were most frequently decentralised to the workplace level in private manufacturing industry. Nevertheless, stewards consulted full-time officers in 38 per cent of cases in that sector. It was in private services that such consultations were least frequent.

Secondly, it is clear from Table V.13 that stewards were more likely to consult full-time officers the larger the size of the workplace. This trend was a little surprising initially, as decentralisation to the workplace level was generally more common in larger workplaces and shop stewards at larger workplaces might be expected to have more resources to deal with issues independently than counterparts at smaller places. Doubtless the greater tendency for stewards at larger establishments to seek outside advice reflected the generally greater union activity at larger workplaces. Stewards from general and industrial unions such as the Transport and General Workers Union and the General, Municipal and Boilermakers

Union were more likely to consult full-time officers than those from craft unions such as the Amalgamated Engineering Union and the Electrical Engineering Telecommunications and Plumbing Union. This pattern was again unexpected, as representatives of a particular craft might have been expected to pay more regard to the implications of technical change for their craft across all workplaces, while semi-skilled or unskilled workers might be expected to pay more attention simply to the implications of change for their own jobs at their own workplaces.

Where manual stewards did consult full-time officers they were generally pleased with the help or advice they received (see Table V.14). In

Table V.14 Manual shop stewards' evaluations of the help received from full-time union officers over the introduction of advanced technical change

Column percentages

	Total	Number of manual workers employed Under 200	200–499	500 or more
Very useful	51	(54)[a]	(42)	51
Fairly useful	28	(24)	(43)	33
Mixed feelings/not stated	10	(8)	(9)	3
Not very useful	7	(8)	(6)	9
Not at all useful	4	(6)	–	4
Mean score[b]	+130	+120	+130	+130
Base: manual shop stewards who consulted full-time officers about advanced technical change				
Unweighted	*138*	*37*	*37*	*64*
Weighted	*42*	*29*	*9*	*5*

[a] See note B.
[b] See note I.

half the cases, stewards said that it was very useful and in the large majority of the remaining cases they felt it to be fairly useful. There were few indications of any marked variation in the level of satisfaction with the help provided by officers as between different categories of steward, or certainly none that were significant, in view of the small number of manual stewards reporting advanced technical change in many sub-sectors.

Other sources of advice or information to manual shop stewards

Full-time officers from their own unions were certainly the chief consultants to manual shop stewards over the introduction of advanced technical change. When we asked if they consulted any other body outside the establishment about the change, only 13 per cent said they had done so and the consultants they mentioned covered a wide span of different types of body. Most frequently they cited educational establishments and official government advisory bodies as their other sources of help or information, but these were mentioned by only two per cent of manual stewards in each case.

The comparative picture among non-manual stewards

It is sometimes suggested that non-manual trade union members and their workplace representatives expect and demand a better service from their union and its full-time officers than manual trade union members and stewards. Non-manual members and stewards, it is said, tend to be more accustomed to dealing with consultants and with people who provide professional services of different kinds, such as accountants, auditors, solicitors and architects. They expect the same kind of service from their trade union as they receive from other professional services. Moreover, the argument continues, non-manual members and representatives, being better informed, tend to think of more questions for their union, through its full-time officers, to answer. Certainly, it is along these lines that some non-manual unions tend to justify the higher dues that they levy from their members, compared with manual unions. Certainly, too, some non-manual unions used the prospect of the impending industrial revolution in offices as a promotional feature in their membership drives(7). The implication was that, if office workers got the security of trade union membership around them, they would be protected from any damaging consequences associated with the introduction of advanced technology.

In view of these arguments, we were rather surprised by our findings on the extent to which non-manual stewards consulted full-time officers about major changes that they had recently experienced, compared with the pattern for manual workers. It is clear from Table V.15 compared with Table V.11, that non-manual stewards were substantially less likely to have consulted full-time officers over the three forms of major change. The contrast was especially marked in relation to the introduction of both conventional technical change and organisational change. It was much less marked in relation to advanced technical change, but it remained true that non-manual stewards were less likely to call upon the services or resources of the union outside the workplace. Nevertheless, the pattern relating to advanced technology was closer to that for organisational change among non-manual stewards than it was among their manual counterparts.

Table V.16 shows that the tendency for stewards in nationalised in-

Table V.15 Extent to which non-manual shop stewards consulted full-time officers over the introduction of different forms of change

Column percentages[a]

	Total	Advanced technical change	Conventional technical change	Organisational change
Full-time officer consulted	36	37	19	42
No consultations	62	62	81	55
Not stated	1	1	–	1
Level of contact[b]				
Local	11	10	4	15
Regional	17	19	6	14
National	15	14	11	22
Base: non-manual shop stewards who reported a major change in the previous three years				
Unweighted	*665*	*520*	*58*	*87*
Weighted	*297*	*201*	*31*	*65*

[a] See note C.
[b] See note F.

Table V.16 Extent to which non-manual shop stewards consulted full-time union officers over the introduction of advanced technical change

Column percentages

	Total	Private manufacturing	Private services	Nationalised industries	Public services
Full-time officer consulted	37	29	33	69	33
No consultations	62	70	65	31	66
Not stated	1	1	2	–	1
Level of contact[a]					
Local	10	13	3	15	11
Regional	19	10	15	47	17
National	14	9	17	26	11
Base: non-manual shop stewards who reported an advanced technical change					
Unweighted	*520*	*169*	*66*	*65*	*191*
Weighted	*201*	*33*	*36*	*22*	*101*

[a] See note F.

Table V.17 Extent to which non-manual shop stewards consulted full-time union officers over the introduction of advanced technical change in relation to size

Column percentages[a]

		Number of manual workers employed at establishment				
	Total	under 50	50-99	100-199	200-499	500 or more
Full-time officer consulted	37	29	35	38	53	48
No consultations	62	71	63	60	46	50
Not stated	1	–	2	2	2	2
Level of contact[b]						
Local	10	11	4	16	11	15
Regional	19	13	23	14	33	24
National	14	16	11	13	10	19
Base: non-manual shop stewards who reported and advanced technical change						
Unweighted	*520*	*56*	*70*	*107*	*135*	*152*
Weighted	*201*	*69*	*50*	*39*	*28*	*15*

[a] See note C.
[b] See note F.

dustries to consult full-time officers very much more than those in any other sector was even more pronounced among non-manual stewards than among their manual counterparts. Private services was the other sector in which non-manual stewards were more active in contacting union offices. Table V.17 shows that there was a similar tendency for non-manual stewards to contact full-time officers more frequently the larger the workplace, as was apparent among manual stewards. Any differences that we were able to identify between the patterns characterising different individual unions appeared to owe more to the characteristic sector in which the union operated than to anything else.

If the indications were that non-manual stewards called upon full-time officers less frequently than their manual counterparts in circumstances of major change, there was some sign that the non-manual grade constituted a more demanding clientele (see Table V.18). Although non-manual stewards generally felt that the response of full-time officers was useful, fewer rated it as very useful. Muted enthusiasm was especially marked among non-manual stewards in smaller workplaces.

Table V.18 Non-manual shop stewards' evaluations of the help received from full-time union officers over the introduction of advanced technical change

Column percentages[a]

	Total	Number of non-manual workers employed		
		Under 100	100–499	500 or more
Very useful	40	(28)[b]	50	56
Fairly useful	37	(44)	30	33
Mixed feelings	10	(16)	3	6
Not very useful	10	(9)	13	5
Not at all useful	3	(2)	5	*
Mean score[c]	+110	+100	+110	+150
Base: non-manual shop stewards who consulted full-time officers about advanced technical change				
Unweighted	*226*	*41*	*112*	*73*
Weighted	*74*	*37*	*30*	*7*

[a] See note C.
[b] See note B.
[c] See note I.

Other sources of advice to non-manual stewards

Non-manual stewards consulted outside bodies other than the full-time officers of their own unions very slightly more frequently than manual counterparts; 17 per cent compared with 13 per cent. The bodies or individuals consulted were varied and numerous and there were few in any particular category. It did appear, however, that non-manual stewards were slightly more likely than manual counterparts to consult convenors or stewards at other establishments. Apart from this tendency, the bodies most frequently contacted were again educational establishments and government advisory bodies, though each was contacted by only two per cent of non-manual stewards confronted with advanced technical change in the office.

Works managers' accounts of the role of personnel managers

Studies of the introduction of the first generation of mainframe computer operations into offices were struck by how infrequently the personnel function was involved in the introduction of the change(8). Managements tended to conceive of the problems in purely technical terms relating to the computer hardware or software or to the analysis of the systems. They rarely recognised explicitly that there might be human considerations relating to a range of decisions associated with the change, to which professional personnel managers could usefully contribute. We did not have time in our main interview with principal management respondents to ask about the role of personnel managers in any major changes that had taken place at the establishment. But in our interviews with works managers in manufacturing industry we were able to pay some attention to that role. We sought interviews with works managers in only those cases where the job title of our principal management respondent included personnel or industrial relations. We knew therefore that the personnel function was represented at the establishment in all cases when we spoke to works managers. On the other hand, we need to stress that the findings in this section on the role of personnel management in the introduction of major change do refer to a minority of the workplaces in our study. First, only one fifth of all the workplaces in the study had personnel managers, but as they were more common in the larger workplaces, 44 per cent of manual workers and 50 per cent of non-manual workers were employed at places where the personnel function was represented. Secondly, our interviews with works managers were necessarily confined to the manufacturing sector.

In view of the extent to which our information on the role of professional personnel managers in the introduction of major change was derived from those workplaces where personnel management had most influence, a very modest picture emerged of the part played by personnel managers. In cases where we interviewed works managers, the personnel department was involved in the introduction in slightly less than one half of the major changes identified (see Table V.19). And even in the instances where the personnel department was involved, it was rare for it to be fully involved from the start. In only 14 per cent of cases did works managers report that personnel management was involved in the initial decision to make the change. The most common pattern, where the personnel department was involved, was for it to be brought in after the decision to make the change had been taken, when its role was to deal with the personnel implications of the change. In only one third of all cases, however, was the personnel department involved either from the start (14 per cent) or immediately after the decision to make the change was taken (20 per cent). The failure to involve personnel managers from the start was even more surprising in the light of the clear evidence of the benefits to be derived from doing so.

Once again our most striking and illuminating source of variation was provided by the three different forms of major change identified (see Table V.19). The personnel department was most heavily involved in

Table V.19 Works managers' accounts of the role of the personnel department in the introduction of major change

Column percentages[a]

	Total	Advanced technical change	Conventional technical change	Organisational change
Personnel department involved	46	50	(13)[b]	(80)
Personnel not involved	52	46	(87)	(20)
Not stated	2	4	(1)	–
Stage of involvement				
Decision to change	14	15	(1)	(30)
Immediately after decision to change	20	19	(2)	(50)
After decision to tell workers	6	9	(2)	–
Later stage	6	7	(8)	–
Base: works managers who reported a major change in the previous three years				
Unweighted	*241*	*176*	*40*	*25*
Weighted	*56*	*37*	*12*	*7*

[a] See note C.
[b] See note B.

organisational change, introduced independently of new plant or machinery. It was involved in the large majority of such cases, and where it was involved it was invariably brought in at an early stage. As we show in Chapter IX, it appeared that many of these organisational changes in manufacturing industry were classic productivity agreements. Since many of them also involved the relaxation of traditional demarcations as between different categories of manual worker, it was hardly surprising that personnel managers were so frequently involved. But we also show in Chapter IX that the introduction of advanced technology was also often associated with the relaxation of those self-same demarcations. Yet personnel managers were very much less frequently involved in the introduction of such changes, and where they were it was often at a much later stage. Perhaps most striking of all was the infrequency with which the personnel department was at all involved in the introduction of conventional technological change. In only 13 per cent of such changes did works managers report any involvement and in most of those instances, personnel managers were not brought in until a comparatively late stage in the proceedings. In terms of broad conclusions on the role of personnel management in the introduction of technical change, it appears that very little has altered since the early 1960s, and that technical change is still largely seen as a technical matter within which there is no established role or function for personnel management.

When we confined our analysis to technical changes and explored patterns of variation in the extent to which the personnel department was involved in the introduction of such changes, we found two striking sources of variation. First, and very importantly, it was clear that where personnel managers were involved, and especially where they were involved at an early stage, the reactions of workers to advanced technical change in manufacturing industry was much more favourable (see Table V.20). The associations shown do not necessarily mean that the involve-

Table V.20 The involvement of personnel management in the introduction of advanced technical change and worker reactions to change

Column percentages

	Worker reactions to change		
	Strongly in favour	Slightly in favour	Some resistance
Personnel management involved	(55)[a]	48	45
Personnel not involved	(42)	44	50
Not stated	(3)	8	5
Stage of involvement[b]			
Involved in decision to make change	(51)	28	8
Immediately after decision to make change	(27)	55	63
When it was decided to tell workers or their representatives	(12)	17	24
At later stage	(9)	*	5
Base: works managers who reported advanced technical change			
Unweighted	*48*	*54*	*54*
Weighted	*14*	*10*	*8*

[a] See Note B.
[b] Base: cases where personnel managers were involved.

ment of personnel managers at an early stage directly resulted in the stronger support for the change. There could be a number of other explanations. On the other hand, the relationship was very strong and consistent. Moreover, our earlier analysis suggested that there was good reason to suppose that there was some tendency to involve personnel managers in change when industrial relations or human problems were felt to be more likely. When, in such circumstances, the involvement of personnel managers emerged as strongly associated with support for the change, then the possibility of a causal link takes on greater plausibility. Secondly, when differences in the size of workplaces were taken into account, personnel departments were more likely to be involved in plants belonging to foreign-owned companies than in plants belonging to domestic companies, and the contrast was marked.

In cases where personnel managers were involved in the introduction of advanced technical change, their main roles, according to works managers, tended to be in reviewing the implications of the change for manning; engaging in consultations or negotiations over the change; advising on the method for introducing the change; and providing a training service.

Footnotes

(1) As mentioned in the Introduction, comments along these lines to ESRC led the Council to sponsor a Company Level Survey to explore the issue of the extent to which workplace surveys created an impression of excessive local autonomy.

(2) Certainly, the indications from the ESRC Company Level Survey had been that workplace respondents attribute more importance to workplace bargaining in pay determination than do managers at higher levels in the organisation.

(3) W.W. Daniel and Neil Millward, *Workplace Industrial Relations in Britain*, Heinemann, London, 1983.

(4) *Ibid.*, and W.W Daniel, *Wage Determination in Industry*, PEP, No.563, 1976.

(5) Michael Fogarty, *Trade Unions and British Industrial Development*, PSI, 1986.

(6) Daniel, 1976, *op.cit.*

(7) Jim Northcott, Michael Fogarty and Malcolm Trevor, *Chips and Jobs: Acceptance of New Technology at Work*, PSI, No.648, 1985.

(8) Enid Mumford, *Computers, Planning and Personnel Management*, Institute of Personnel Management, 1969.

VI Worker and Trade Union Involvement in the Introduction of Advanced Technical Change

One point about which all parties and representative groups in industry, commerce and administration agree is that there should be worker involvement in the introduction of technical change. The Confederation of British Industry, the British Institute of Management, the Trades Union Congress, individual trade unions, and the major political parties all agree that worker involvement in decision-making is generally desirable(1). They agree, too, that it is particularly desirable when new technology is being introduced. There may be disagreement over the preferred medium for such involvement; whether, for instance, it should be through collective bargaining, joint consultation, joint decision-making, codetermination or shop floor participation, or a particular mixture of media. There may be disagreement over the rationale for worker participation; whether, for example, it rests upon workers' rights independently of any practical implications or upon more pragmatic concerns with winning consent for change, improving the quality of decisions or accommodating to powerful groups in a position to obstruct change. There may be disagreement over whether the arrangements for worker involvement should be voluntary or statutory, but there is complete accord upon the desirability of involvement. In this the parties have the support of a strong body of research evidence. A thorough review of this evidence came to the following conclusion, so widely quoted that it has become almost a cliche:

> There is hardly a study in the entire literature which fails to demonstrate that satisfaction in work is enhanced or that other generally acknowledged beneficial consequences accrue from a genuine increase in workers' decision-making power. Such consistency of findings, I submit, is rare in social research(2).

Such conclusions and the evidence supporting them have generally been a staple part of the diets of management courses in colleges, universities and business schools, since the classic human relations studies of the 1930s and 1940s(3).

In relation to this large, united and apparently overwhelming army, the only people who, according to our present findings, are out of step are the practising managers upon whom the implementation of the arrangements for worker involvement depends. Our findings show that, in many instances, there was little or no involvement of workers and their representatives in the introduction of the major changes we studied. The overall pattern which emerged revealed that managers generally operated in a minimally pragmatic fashion. If they could introduce the changes they wanted without having to take account of the views of any other group or individual, they did so. Where they consulted or negotiated, it was principally because they were required to do so, either by the industrial relations institutions at their workplaces, or by resistance to the changes they wanted on the part of workers or their representatives. There was no hint of managerial commitment to worker involvement as a means of improving the form of the change or generating enthusiasm for it. The fact that there was, nevertheless, as we show in the next chapter, so much support from workers for advanced technical change speaks volumes for the widespread, spontaneous appetite for technical change among British workers. This summary of the extent of worker involvement in technical change applied particularly strongly to change affecting manual workers. There did appear to be more inclination on the part of managers to consult non-manual workers or their representatives about office change. It appeared that this was because managers found it easier to talk with office workers about change, were less fearful of unhelpful reactions or more hopeful of helpful ones. It was certainly not because better machinery existed for the involvement of office staff.

When the implications of microelectronic technology for change in industry and offices first started to become apparent to commentators in Britain, the focus of attention in relation to worker involvement was the desirability of new technology agreements(4). These were intended to provide a comprehensive framework within which to introduce advanced technology on an agreed and harmonious basis. Initial attention focussed upon how frequently such agreements were introduced and what they contained. Soon, however, it became apparent that the drive for new technology agreements lost impetus(5). In consequence, we did not concern ourselves with the extent or content of any such procedural agreements. We focussed upon what actually happened in practice, so far as consultation and negotiation were concerned, when particular major changes were introduced.

First, we took as our model the types of arrangement for consultation and participation that have been adopted by companies with well-developed personnel policies and practices who wanted to introduce major change(6). These have included discussions with joint shop stewards' committees about the proposal to introduce the change and its implications; the establishment of a joint working party with worker representation to co-ordinate the implementation of the change; the setting up of discussion groups among the workers directly affected by the change; a range of informal sets of discussion; and the further discussion of the final package arrived at with the joint shop stewards' committee. We by no means assumed that managements should have set up such arrangements, but the model provided us with a range of forms of consultation in which managers have engaged when introducing major change and which have been widely written about and discussed, certainly since the 1960s. We asked managers whether they had had discussions or consultations of the following types about the introduction of their most recent advanced technological change:

- informal discussions with individual manual workers;
- meetings with groups of manual workers;
- discussions in an established joint consultative committee;
- discussions in a specially constituted committee set up to consider the change;
- discussions with union representatives at the establishment;
- discussions with union officials outside the establishment.

We asked a parallel set of questions about the involvement of office workers in technical change in the office. In practice, of course, there are a whole series of stages in the introduction of advanced technical change when there may be involvement in decisions by workers or their representatives. These include the decisions to invest in new plant or machinery; the decisions about which equipment to buy; and all the personnel, technical and practical decisions associated with the installation of the new technology and getting it working. It was not possible in an inquiry such as ours to explore the extent of involvement at each of these stages and in each of the decisions associated with the different stages. It will be noted that, so far as our management respondents were concerned, we essentially asked about discussions in relation to any decision at any stage. So far as shop stewards were concerned, we were able, as there was less pressure on time in our interviews, to distinguish between two stages and ask separately about discussions concerned with, first, *whether* the change should be introduced and, secondly, *how* the change should be implemented.

Our questions about consultations and discussions were essentially derived from a view of worker involvement in change based upon an enlightened self-interested managerial perspective. In addition to these questions we also sought, secondly, to establish how far the change was subject to joint regulation and thereby provided the form of worker involvement sought by the trade union movement(7). That is to say, we asked whether the introduction of the change was . . .

- negotiated with union representatives and dependent upon their agreement, or
- discussed with union representatives in a way which took their views into account but left management free to make the decisions, or
- not discussed with union representatives.

As was our general practice, the questions about consultations and negotiations were asked separately with regard to advanced technical changes affecting manual workers and office workers. They were also asked of both managers and shop stewards.

Some details of the advanced technical changes used for evaluating the extent of consultation and negotiation

In order to place our findings into context it is useful at this point to summarise the different features of the changes which demonstrate how substantial they were. They generally affected about one quarter of manual workers at the workplaces where they were introduced. These places were generally large and in most cases employed more than 300 manual workers. The changes frequently affected earnings and manning levels. The technology involved was relatively new and unfamiliar. In a majority of cases the new plant or equipment changed key features of workers' jobs, including the skill they exercised, the range of the tasks they did, their control over the job and the interest furnished by the work. In short, the extent to which there were consultations or negotiations over these changes provided a touchstone of the extent to which workers and their representatives had the opportunity to influence major technical changes affecting their jobs at their place of work.

The extent of consultations over changes affecting manual workers

Table VI.1 summarises managers' answers to our questions on whether discussions or consultations of any of the listed types were held at any stage of the change. It is immediately apparent that, even according to the accounts of managers, the extent of different forms of consultation was modest. In nearly one fifth of cases managers reported that they had no consultations of the types listed. They spoke with individual workers in only a modest majority of cases. Shop stewards were involved in discussions of the introduction of major technical changes in only 39 per cent of cases. Of course, manual unions were not recognised in all work-

Table VI.1 Extent and form of consultation over the introduction of advanced technical change in relation to number of manual workers employed

Column percentages[a]

	All establish-ments	Number of manual workers at establishment 25-49	50-99	100-199	200-499	500–999	1000 or more	All manual employees
One or more of the listed forms of consultation	82	70	89	91	90	99	97	92
No consultation	17	28	9	7	10	2	–	7
Not stated	1	2	1	1	–	*	3	1
Formal — union channels[b]								
Discussed with shop stewards	39	22	41	47	66	83	88	66
Discussed with full-time officers	16	17	11	11	21	25	29	23
Formal — non-union channels								
Discussed in established JCC	17	11	11	23	26	48	55	35
Discussed in specially constituted committee	11	11	9	5	14	25	28	18
Informal								
Discussions with individual workers	58	56	56	62	61	61	65	63
Meetings with groups of workers	38	29	40	39	50	69	68	53
Base: establishments with 25 or more manual workers and experiencing advanced technical change								
Unweighted	*458*	*50*	*56*	*70*	*121*	*104*	*57*	
Weighted	*212*	*91*	*51*	*31*	*27*	*8*	*4*	

[a] See note C.

places included in the table. But even when we confined analysis to places where manual unions were recognised, the proportion of cases which involved consultations with stewards rose to only 52 per cent (see Table VI.4). Moreover, these figures were based upon the reports of managers. When we compare them later in the chapter with the accounts of stewards there emerges an impression of even less consultation.

It is true, first, as Table VI.1 shows, that all forms of consultation we listed were more common the larger the size of the workplace. In consequence, the proportion of manual workers employed at places where there were no consultations was substantially lower than the proportion of workplaces where that was the case. The tendency for consultations to be more common in larger establishments was most marked for formal channels of consultation, such as joint consultative committees, and discussions with shop stewards and full-time officers. But, more surprisingly, informal channels, such as discussions with individual workers and meetings with groups of workers, were also slightly more frequently used in larger workplaces.

Secondly, consultations of all types were very much less common in the private service sector than in private manufacturing or the public sector (see Table VI.2). This remained true even when differences in size were taken into account. In larger private service establishments (employing 200 or more manual workers), managers reported no consultations over the introduction of advanced technical change in 46 per cent of cases (see Table VI.3). The pattern in the public sector was strongly influenced by the nationalised industries where consultations through formal channels were most pronounced. It may appear that in public services the extent of consultations through formal channels was remarkably low compared with private manufacturing industry. This difference resulted, however, from the characteristic size of workplaces in the two sectors. When we confined our analysis to the larger workplaces it was clear that consultation through all formal channels was more common in public services.

Variations in levels of consultation in relation to trade union organisation and extent of support for change

The most striking features of the pattern of variation that we identified in relation to forms and levels of consultation with manual workers and their representatives related to level of trade union organisation at the workplace and level of support for the change. First, our analysis showed a strong tendency for levels of consultation to increase the higher the level of unionisation at workplaces. It was not simply that, as might be expected, discussions with both shop stewards and full-time officers were more common in such workplaces. Where manual union organisation was strong, informal consultations with individuals and groups were also more common. Secondly, there were more consulations of all types where

Table VI.2 Extent and form of consultation over introduction of advanced technical change in relation to sector

Column percentages[a]

	All establishments	Private manufacturing	Private services	Nationalised industries	Public services
One or more of the listed forms of consultation	82	85	64	100	83
No consultation	17	13	37	–	17
Not stated	1	2	–	*	–
Formal—union channels[b]					
Discussed with shop stewards	39	43	19	60	37
Discussed with full-time officers	16	12	5	52	12
Formal—non-union channels					
Discussed in established JCC	17	18	9	32	10
Discussed in specially constituted committee	11	12	5	24	4
Informal					
Discussions with individual workers	58	64	51	47	55
Meetings with groups of workers	38	38	26	55	43
Base: establishments with 25 or more manual workers and experiencing advanced technical change					
Unweighted	*458*	*286*	*59*	*63*	*50*
Weighted	*212*	*116*	*44*	*26*	*28*

[a] See note C.
[b] See note F.

there was less support for change. In combination, as we suggested at the start of this chapter, the pattern implied that managers generally tended to consult over the introduction of change affecting manual workers when they were required to do so either by the existence of trade union organisation or by reluctance to accept change among workers or stewards.

So far as trade union organisation was concerned, there were major differences in the extent of consultation where trade unions were recognised compared with the places where they were not (see Table VI.4). First, in 44 per cent of places where manual workers were not represented by a union managers reported that there was no consultation

Table VI.3 Consultation over advanced technical change in relation to size and sector

Column percentages

	Large establishments (employing 200 or more manual workers)		
	Private manufacturing	Private services	Public services
One or more of the listed forms of consultation	96	(55)[a]	(93)
No consultation	4	(45)	(7)
Not stated	*	–	–
Formal—union channels[b]			
Discussed with shop stewards	71	(30)	(87)
Discussed with full-time officers	23	(6)	(46)
Formal—non-union channels			
Discussed in established JCC	31	(31)	(68)
Discussed in specially constituted committee	18	(6)	(39)
Informal			
Discussions with individual workers	65	(50)	(52)
Meetings with groups of workers	55	(34)	(63)
Base: establishments with 25 or more manual workers and experiencing advanced technical change			
Unweighted	*191*	*23*	*23*
Weighted	*25*	*4*	*3*

[a] See note B.
[b] See note F.

of the types listed. This proportion compared with eight per cent in places where unions were recognised. Secondly, every type of consultation was less common where manual unions were not recognised, and this was true for informal as well as the more formal channels. In general, employers who do not recognise unions tend to say that they prefer to treat employees as individuals rather than as part of a collective(8). Occasionally they have promoted joint consultative committees as a substitute for trade union representation(9). But our present findings show that, in

Table VI.4 Extent and form of consultations over the introduction of advanced technical change in relation to union recognition

Column percentages[a]

		All establishments		Establishments with 50–199 manual workers	
	All establish-ments	Manual union recognised	Union not recognised	Manual union recognised	Union not recognised
One or more of the forms of consultation	82	91	54	94	(75)
No consultation	17	8	44	5	(21)[b]
Not stated	1	1	1	1	(4)
Formal — union channels[c]					
Discussed with shop stewards	39	52	–	56	–
Discussed with full-time officers	16	21	–	14	–
Formal — non-union channels					
Discussed in established JCC	17	20	5	16	(12)
Discussed in specially constituted committees	11	13	4	9	(2)
Informal					
Discussions with individual workers	58	62	45	61	(51)
Meetings with groups of workers	38	42	25	39	(42)
Base: establishments with 25 or more manual workers and experiencing advanced technical change					
Unweighted	*458*	*405*	*53*	*100*	*26*
Weighted	*212*	*160*	*52*	*64*	*18*

[a] See note C.
[b] See note B.
[c] See note F.

practice, the absence of discussions with shop stewards or full-time union officers was not replaced in non-union shops by any other medium – neither joint consultative machinery through elected representatives, nor special *ad hoc* groups, nor even talking directly with the workers involved either individually or in groups. Thirdly, it was not simply that our contrast between unionised and non-union workplaces compared larger workplaces where more formal consultations might be appropriate with small units where it might be less so. All the establishments in the analysis employed 25 or more manual workers. When we compared workplaces employing from 50 to 199 manual workers, as in the second part of Table VI.4, the main features of the contrast between union and non-union workplaces remained, though they became less pronounced. Fourthly, in places where there were union members, there was a tendency for all channels of consultation to be used more frequently the higher the level of trade union density (see Table VI.5).

We analyse levels of workers' and trade union support for change in Chapter X. We establish through that analysis that, according to the accounts of both managers and shop stewards, there was strong and widespread support for advanced technical change. It may well be that this was one of the main reasons for low levels of consultation, because we also established that in the minority of places where there was resistance the level of consultation rose substantially. The association between resistance to change and increased levels of consultation was especially strong so far as the reactions of stewards were concerned. It is implausible that the initiation of consultations by managers actively encouraged resistance and much more plausible that consultations were a response to initial reluctance on the part of stewards or workers to accept the change. The association between unfavourable reactions and negotiation over the change was even stronger than the association with consultation. In combination, our results showing that both unionisation and resistance to change among manual workers were associated with increased consultation over change suggested that the initiative for worker involvement in change came very much from workers and their representatives. Lest anyone should feel that this will be inevitably and invariably the case, it should be noted that our analysis of consultation over change in the office, later in this chapter, shows a different pattern.

Levels of negotiation over change

As we indicated at the start of the chapter, when commentators first began to discuss the implications of the microelectronic revolution they tended to see comprehensive technological change agreements as forming the spearhead for introducing change. The assumption was that it would be necessary and desirable for advanced technical change to be introduced on the basis of joint regulation if it was going to be acceptable.

Table VI.5 Consultation over advanced technical change in relation to union density

Column percentages[a]

	Manual union density		
	1-49 per cent	50-59 per cent	100 per cent
One or more of the listed forms of consultation	(68)[b]	87	97
No consultation	(32)	13	3
Not stated	–	–	*
Formal – union channels[c]			
Discussed with shop stewards	(17)	51	58
Discussed with full-time officers	(1)	15	29
Formal – non-union channels			
Discussed in established JCC	(15)	15	25
Discussed in specially constituted committee	(21)	210	12
Informal			
Discussions with individual workers	(54)	55	69
Meetings with groups of workers	(38)	42	46
Base: establishments with 25 or more manual workers and experiencing advanced technical change			
Unweighted	*30*	*147*	*197*
Weighted	*26*	*62*	*75*

[a] See note C.
[b] See note B.
[c] See note F.

This analysis followed on predictions of the increasing need for management by agreement in the 1970s(9).

In contrast to such predictions, the results already reported in this chapter show that many of the advanced technical changes in our study were introduced without consultation, let alone agreement. And yet the changes were still strongly supported by the workers affected. Our question, outlined in the introduction to this chapter, which sought to establish how far the introduction of major technical change was subject to joint regulation, revealed that this was rarely the case. It might very well have been predicted from the analysis of the extent of consultation. Nevertheless, it still came as a surprise to find, according to management accounts, that even when analysis was confined to places recognising unions, changes were negotiated in fewer than one out of every ten cases (see Table VI.6). Such cases represented six per cent of all establishments that had recently introduced advanced technical change affecting manual workers. In principle, it is possible that one reason why negotiated change was so rare, was because places that needed trade union agreement in order to introduce change had failed to introduce new technology. In practice, this possibility was wholly inconsistent with our general analysis of the implications for change of a high level of trade union organisation. In particular, we showed in Chapter II that there was no tendency for major change to have been less common in places where union organisation was strong than in those where it was weak or there was no union organisation.

It is true that negotiated change was most common in manufacturing industry and was more frequent the larger the size of the manual workforce. But even in manufacturing establishments employing 500 or more manual workers, the proportion where the change was negotiated rose to only 16 per cent. In addition to, and independently of, the number of manual workers employed, the level of manual union density also had a strong influence upon the extent of joint regulation. In large workplaces with 100 per cent manual union membership – a *de facto* closed shop – the proportion of workplaces where the advanced technical change was introduced through negotiation rose to 27 per cent. Thus, even in circumstances most favourable to the development of comprehensive technological change agreements, the practice of negotiating the introduction of technical change achieved a penetration of only about one quarter of workplaces.

Manual stewards' accounts of levels of consultation and negotiation

So far our analysis of levels and forms of consultation and negotiation over the introduction of advanced technical change affecting manual workers has been based upon the reports of managers. As was generally the case with regard to our survey design in this study, the addition of the

Table VI.6 Extent of negotiations over advanced technical change in relation to number of manual workers employed

Column percentages[a]

		Number of manual workers					
	All establishments	25-99	100-199	200-499	500-999	1000 or more	All manual employees
Change negotiated with union representatives	8	5	10	14	17	15	13
Consultation with union representatives	61	58	61	68	75	76	69
No discussions with union representatives	30	37	27	18	7	3	16
Not stated	1	–	2	–	1	5	2
Base: establishments with 25 or more manual workers, recognising manual unions and experiencing advanced technical change							
Unweighted	*389*	*73*	*56*	*108*	*98*	*54*	
Weighted	*154*	*95*	*25*	*24*	*8*	*4*	

[a] See note C.

accounts of shop stewards to this picture added two important features to the analysis. First, it enabled us to present a more balanced account, since it was clear that, in many instances, the differences in perspective between managers and shop stewards resulted in differences in detail emerging from their respective reports of the same or very similar events. Secondly, we were able to ask a little more about different aspects of the introduction of advanced technical change in our interview schedule with shop stewards. Accordingly, as we explained at the beginning of this chapter, we were able to focus the questions we asked shop stewards about consultation upon two distinct contexts: first, the decision to introduce the change and, secondly, decisions about how to introduce the change. In Table VI.7, we combine the answers of stewards to the two separate questions into one column. That is to say, we found that stewards reported that there were discussions with manual stewards over the introduction of advanced technical change, either before the change was introduced or during its implementation, in 84 per cent of cases where manual unions were recognised. They reported that there was no consultation at either stage in 14 per cent of cases.

Our analysis of managers' reports revealed that levels of consultation were remarkably low, especially in view of the very high level of consensus about the desirability of consulting with workers or their representatives over the introduction of change. We found, too, that managers tended to consult only when they were required to do so, in response to either trade union structures or initial resistance from workers or their representatives. This meant that the cases where we were able to compare reports from both managers and stewards were ones in which the level of consultation was relatively high. In other words, they were confined to workplaces where manual unions were recognised and there were manual stewards at the workplace. It is clear from Table VI.7 that stewards were inclined to report even less consultation than managers. This was especially the case when analysis was confined to workplaces where we interviewed managers and manual stewards and both reported on levels and forms of consultation. Stewards reported every type of consultation less frequently than managers, other than the setting up of special *ad hoc* consultative committees as a forum for discussing the introduction of the change. They were particularly less likely to report discussions with union officers and individual manual workers, and they were substantially more likely to report that there was no consultation at all.

Table VI.8 adds the refinement of distinguishing between consultations before the introduction of the advanced technology and consultations during its implementation. It is immediately apparent that the overall picture of very modest levels of consultation was further reinforced. When the focus became each of the two distinct contexts, then the proportion of cases where there was no consultation rose. Lack of

Table VI.7 Comparison of managers' and manual shop stewards' accounts of the extent and form of consultation over the introduction of advanced technical change affecting manual workers

Column percentages[a]

	Union recognised			
	All managers reporting advanced technical change	All stewards reporting advanced technical change	Establishments where both were interviewed about advanced technical change: Managers	Stewards
One or more of listed forms of consultation	91	84	95	81
No consultations	8	14	4	17
Not stated	1	2	1	2
Formal—union channels[b]				
Discussed with shop stewards	52	41	59	46
Discussed with full-time officers	21	17	30	18
Formal—non-union channels				
Discussed in established JCC	20	24	29	20
Discussed in specially constituted committee	13	17	21	20
Informal				
Discussion with individual workers	62	49	63	50
Meetings with groups of workers	42	39	49	42
Base: see footnote[b]				
Unweighted	*405*	*288*	*205*	*205*
Weighted	*160*	*101*	*72*	*72*

[a] See note C.
[b] See note F.
[c] In the first two columns we compare the accounts of all managers who reported advanced technical change in places where manual unions were recognised with the reports of all the manual shop stewards who reported such change. In the second two coloumns we confine the analysis to cases where managers and stewards both reported advanced technical change. The first contrast shows the picture that would have been revealed had we depended solely upon the accounts of managers, compared with the results had our only source of information been the reports of stewards. The second contrast highlights differences between management and union perspectives, perceptions or recollections when talking about the same events in the same workplace.

Table VI.8 Manual shop stewards' accounts of consultations over the introduction of advanced technical change before the decision and during implementation

Column percentages[a]

	All establishments		Private manufacturing		Private services		Nationalised industries		Public services	
	Before	During	Before	During	Before	During	Before	During	Before	During
One or more of listed forms of consultation	60	75	71	70	(70)[b]	(83)	(37)	(88)	(45)	(59)
No consultations	39	22	27	26	(30)	(17)	(63)	(12)	(55)	(33)
Not stated	1	3	2	3	–	*	(1)	(1)	*	(8)
Formal — union channels[c]										
Discussed with shop stewards	27	33	34	29	(11)	(23)	(27)	(45)	(20)	(40)
Discussed with full-time officers	12	12	11	8	(2)	(14)	(18)	(17)	(13)	(16)
Formal — non-union channels										
Discussed in established JCC	17	21	16	17	(5)	(13)	(27)	(39)	(14)	(12)
Discussed in specially constituted committee	11	14	17	17	(1)	(11)	(5)	(14)	(7)	(10)
Informal										
Discussions with individual workers	29	40	33	38	(51)	(59)	(7)	(39)	(23)	(27)
Meetings with groups of workers	23	25	33	23	(16)	(22)	(10)	(33)	(18)	(26)
Base: manual stewards who reported advanced technical change										
Unweighted	*288*	*288*	*166*	*166*	*29*	*29*	*49*	*49*	*44*	*44*
Weighted	*101*	*101*	*49*	*49*	*16*	*16*	*23*	*23*	*14*	*14*

[a] See note C.
[b] See note B.
[c] See note F.

consultation over whether to introduce the change was especially common. There were, however, marked differences between different sectors in this pattern. Nationalised industries revealed the general pattern to the most pronounced degree. We noted earlier, when analysing managers' accounts, that managers of nationalised industries were less likely than in any other sector to report that there had been none of the listed types of consultation over the change. The accounts of stewards confirmed that lack of consultation over the *implementation of change* was least common in nationalised industries. But it was more common than in any other sector for no-one at workplace level to be consulted over the prior decision about *whether* the change should be introduced. In over two thirds of cases, stewards in nationalised industries reported that there were no consultations of any of the listed types before it was decided to introduce the change. In private services, too, there was a marked tendency for any consultations to be more common at the implementation stage.

But manufacturing industry was the dominant sector so far as the introduction of advanced technical change affecting manual workers was concerned. In that sector the levels of consultation reported by stewards at the two stages were remarkably similar. It is noteworthy that they reported no consultation over whether to introduce the change in one quarter of cases and also no consultation over the implementation of the change in one quarter of cases. Of course, this pattern relates to the parts of manufacturing industry where manual unions were recognised and where consultation over technical change was relatively common. Moreover, the fact that, according to the accounts of stewards themselves, there were consultations with stewards before the introduction of one third of the changes in manufacturing and also consultations over implementation in nearly one third of cases, did not mean that there were consultations *both* before and during the change in that proportion of cases. In some instances, there were consultations before but not during; in others, the opposite was true. When we calculated the proportion of cases where there were particular types of consultation both before the introduction of the change and during its implementation we found the picture shown in Table VI.9. The proportion of cases where stewards reported that they or colleagues were consulted at both stages fell to one fifth. Levels of formal consultation in private services emerged as relatively very low indeed according to this more comprehensive definition.

Manual stewards were no more likely than managers to report that the introduction of advanced technical change was negotiated. Indeed, their reports of the extent to which the change was subject to joint regulation were very similar to those of managers (see Table VI.10). The pattern in relation to this item, however, was more interesting when we came to compare the reports from non-manual representatives and managers about technical changes in the office. Accordingly, we discuss manage-

Table VI.9 Manual stewards' accounts of consultations both before the introduction of advanced technical change and during its implementation

Column percentages

	Total	Private manufacturing	Private services	Nationalised industries	Local/central government
Formal union channels					
Discussed with shop stewards	19	20	(10[a])	(22)	(20)
Discussed with full-time officers	6	5	(2)	(9)	(7)
Formal non-union channels					
Discussed in established JCC	13	13	(3)	(22)	(10)
Discussed in specially constituted committee	8	13	(1)	(2)	(7)
Informal					
Discussions with individual workers	20	24	(39)	(4)	(8)
Meetings with groups of workers	10	13	(11)	(3)	(7)
Base: manual stewards who reported advanced technical change					
Unweighted	*288*	*166*	*29*	*49*	*44*
Weighted	*101*	*49*	*16*	*23*	*14*

[a] See note B.

Table VI.10 Comparison of managers' and manual stewards' accounts of the extent of negotiations over advanced technical change

Column percentages[a]

	Manual union recognised		Workplaces where both managers and stewards were interviewed about an advanced technical change	
	All managers	All shop stewards	Managers	Stewards
Change negotiated with union representatives	8	13	14	10
Consultation with union representatives	61	59	67	63
No discussions with union representatives	30	23	17	21
Not stated	1	4	2	6
Base: as column heads (a)[b]				
Unweighted	*405*	*288*	*205*	*205*
Weighted	*160*	*101*	*72*	*72*

[a] See note C
[b] See note to Table VI.7

ment and union perspectives upon joint regulation more fully in our section on office changes.

Consultation over advanced technical change in the office

We were struck by how little consultation there was with manual workers and their representatives over the introduction of advanced technical change and how much that consultation depended upon pressures being placed upon managements by trade union institutions or less favourable reactions from the workers affected. We found a different pattern regarding consultations over the introduction of technical change in the office. Some form of consultation was more common and it was less dependent upon pressure from people or institutions. We cannot comment, from the information provided by our present study, on how far this contrast depends upon the traditional social class divisions in Britain and the associated difficulties commonly attributed to British managers in communicating with the members of the manual working class. Whatever the reasons, the differences were marked, as Table VI.11 shows.

The contrast was especially pronounced so far as informal discussions and *ad hoc* arrangements for consultation were concerned. For instance, in places where both manual workers and office workers were affected by advanced technical change, managers reported that they discussed the changes with individual office workers in 82 per cent of cases but with individual manual workers in only 62 per cent of cases. On the other hand, consultations over change affecting office workers were less likely to take place within the framework of trade union representation or established joint consultative machinery. Part of the reason for this contrast, of course, was the fact that unions representing manual workers were more widely recognised. But even when we confined the comparison to places where both manual and non-manual unions were recognised, it remained the case that consultations through union channels were much less common in relation to office change while informal consultations with office workers remained more common (see Table VI.12). Although some form of consultation, and especially informal discussions with individuals or groups, were very much more common among office workers affected by change, there remained a minority of cases of technological change in offices where no consultation took place. Even according to managers, there were, in about one case in every ten, no discussions with individual office workers about the introduction of the change in the office or its implementation, nor with groups of workers nor in a committee either established or *ad hoc*, nor with a union representative, whether shop steward or full-time officer. The accounts of non-manual stewards reported later in the chapter suggested that this proportion was even more substantial. Cases where there were no consultations were concentrated in small workplaces, in private services and in places where no non-manual union was recognised.

Table VI.13 shows the pattern of non-manual consultation according to managers' reports in relation to sector. So far as informal consultations with individuals or groups of office workers were concerned, the differences between the sectors were not marked. Discussions of both kinds were slightly less common than normal in the public services. Nationalised industries were especially likely to report discussions in groups but less likely to report discussions with individuals. This pattern contrasted most strongly with private manufacturing industry, where individual consultations were most frequently reported and group discussions less frequently so.

When it came to formal channels of consultation the contrasts between sectors were more marked. There were discussions over changes affecting office workers in established joint consultative committees very much more frequently in the public sector, generally, and in nationalised industries, particularly. The use of *ad hoc* committees to discuss change was

Table VI.11 Consultations with manual workers over advanced technical change compared with consultations with office workers

Column percentages[a]

	All relevant establishments		Establishments where advanced technical change affected both categories of employee	
	Manual workers	Office workers	Manual workers	Office workers
One or more of listed forms of consultation	82	89	89	94
No consultations	17	9	10	5
Not stated	1	2	1	1
Formal—union channels[b]				
Discussed with shop stewards	39	20	60	29
Discussed with full-time officers	16	9	18	9
Formal—non-union channels				
Discussed in established JCC	17	12	24	14
Discussed in specially constituted committee	11	14	12	14
Informal				
Discussions with individual workers	58	76	62	82
Meetings with groups of workers	38	45	41	45
Base: all relevant establishments[c]				
Unweighted	*458*	*977*	*326*	*326*
Weighted	*212*	*500*	*78*	*78*

[a] See note C.
[b] See note F.
[c] See note to Table V.5.

Table VI.12 Comparison of extent and methods of consultation where both manual and non-manual unions were recognised

Column percentages[a]

	All relevant establishments		Establishments where advanced technical change affected both categories of employee	
	Manual workers	Office workers	Manual workers	Office workers
One or more of listed forms of consultation	90	95	92	95
No consultations	8	4	7	4
Not stated	2	1	1	1
Formal—union channels[b]				
Discussed with shop stewards	53	35	75	46
Discussed with full-time officers	21	16	27	15
Formal—non-union channels				
Discussed in established JCC	19	17	31	21
Discussed in specially constituted committee	13	16	15	17
Informal				
Discussions with individual workers	61	73	58	78
Meetings with groups of workers	42	47	49	51
Base: all relevant establishments where unions were recognised[c]				
Unweighted	*405*	*736*	*269*	*269*
Weighted	*160*	*283*	*48*	*48*

[a] See note C.
[b] See note F.
[c] See note to Table V.5.

Table VI.13 Extent and form of consultation over the introduction of advanced technical change in the office in relation to sector

Column percentages[a]

	Total	Private manu-facturing	Private services	National-ised industries	Public services
One or more of listed forms of consultations	89	90	84	100	96
No consultation	9	8	14	*	3
Not stated	2	2	3	*	1
Formal—union channels[b]					
Discussed with shop stewards	20	16	10	66	35
Discussed with full-time officers	9	4	4	57	16
Formal—non-union channels					
Discussed in established JCC	12	8	8	49	19
Discussed in specially constituted committee	14	11	16	18	12
Informal					
Discussions with individual workers	76	81	75	65	73
Meetings with groups of workers	45	41	48	63	41
Base: establishments with 25 or more non-manual workers that introduced advanced technical change in the office					
Unweighted	*977*	*349*	*255*	*105*	*269*
Weighted	*500*	*121*	*231*	*24*	*124*

[a] See note C.
[b] See note F.

at a similar level in all sectors, but this still meant that, when the use of the two types of committee were combined, discussion in some kind of committee remained much more common in the public sector.

Consultations with union officers over change affecting office workers were also very much more common in the public sector. The difference was striking as regards consultations with both shop stewards and full-time officers but it was most marked in relation to full-time officers. Nationalised industries stood out as the sector where union officers, and especially full-time officers, were most likely to be consulted. In nationalised industries, managers reported that they discussed the change with full-time officers in over one half (57 per cent) of the cases where computers or word processors were introduced into the office. In both parts of the private sector the equivalent proportion was four per cent. Of course, the contrast between the public and private sectors in the extent to which trade union officers were consulted owed much to the fact that non-manual unions were much more likely to be recognised in the public sector. When, however, the comparison was confined to places where unions were recognised the differences were still large. This was especially true so far as discussions with full-time officers were concerned. The difference in the extent of consultations with stewards became less marked. Indeed, the contrast between private manufacturing industry and public services became slight, but nationalised industries remained much more likely than any other sector to engage in consultations with stewards over the introduction of change.

The balance between informal and formal consultations also varied, independently of sectoral differences, in relation to size as measured by the number of non-manual workers employed (see Table VI.14). There were strong and consistent tendencies for discussions through established joint consultative committees and with union officers to be more frequently reported the more non-manual workers were employed. This was true both overall and when analysis was confined to cases where trade unions were recognised. When these trends were analysed in relation to both size and sector, it became apparent that each was a strong independent source of variation in the patterns identified. So far as individual consultations were concerned, there was a mixed pattern in relation to size and a slight, but by no means strong or consistent, tendency for discussions with groups to be more frequently reported the larger the size of the workforce.

So far as other aspects of ownership were concerned, it appeared that consultation through formal methods of all types was very much more common in places belonging to groups than in independent workplaces until a certain size was reached, when the contrast disappeared. More detailed analysis revealed, however, that the apparent contrast between independent workplaces and those that were part of groups arose largely because non-manual unions were so much less frequently recognised at independent workplaces.

It was difficult to identify any consistent differences in the extent to which managers reported consultations through various means in relation to the nationality of the ownership of workplaces. Overall, there appeared to be little difference in either the level of consultation, or the channels. Indeed, it appeared that, when workplaces of similar size were compared, there was a slight tendency for UK workplaces to use each of the channels more frequently, apart from discussions with individuals.

Table VI.14 Extent and form of consultation over the introduction of advanced technical change in the office, in relation to number of non-manual workers employed

Column percentages[a]

	All establishments	25-49	50-99	100-199	200-499	500-999	1000 or more	All non-manual employees
One of more of listed forms of consultation	89	81	95	92	95	96	98	94
No consultations	9	15	4	7	4	4	2	5
Not stated	2	4	1	1	1	–	–	1
Formal — union channels[b]								
Discussed with shop stewards	20	10	22	23	41	48	53	37
Discussed with full-time officers	9	6	8	7	20	28	25	17
Formal — non-union channels								
Discussed in established JCC	12	5	12	13	30	29	42	26
Discussed in specially constituted committee	14	14	13	15	17	19	16	17
Informal								
Discussions with individual workers	76	71	85	72	77	73	84	77
Meetings with groups of workers	45	43	47	40	52	53	55	51
Base: establishments with 25 or more non-manual workers that introduced advanced technical change in the office								
Unweighted	*977*	*129*	*165*	*220*	*223*	*137*	*103*	
Weighted	*500*	*201*	*137*	*91*	*47*	*4*	*10*	

[a] See note C.

This tendency was most marked so far as larger workplaces were concerned.

Consultations in relation to non-manual union organisation

A feature of our findings concerning consultations with manual workers or their representatives over the introduction of technological change was the extent to which discussions of all kinds were less common in places where trade unions were not recognised. The pattern in relation to offices took a similar form but was less marked. As Table VI.15 shows, there was a tendency for managers more frequently to report consultations of all kinds, apart from discussions with individuals, where unions were recognised. But this exception was a major one which very much distinguished the pattern from that for manual workers. Moreover, there was also little difference in the extent to which managers reported discussions with groups of office workers in places where unions were recognised compared with those where they were not. When establishments of similar size were compared, the overall differences between places that recognised unions and those that did not largely persisted, but they were most marked in smaller workplaces, those employing fewer than 50 non-manual workers. In such workplaces there was no consultation over the introduction of word processors or computers in nearly one quarter of cases. So far as union density was concerned, there were strong and consistent tendencies for managers to be more likely to report discussions with both shop stewards and full-time officers the higher the proportion of union members at the workplace, but, again, there was not the same tendency apparent in relation to manual workers. The other striking feature in Table VI.16 was the infrequency with which any method other than discussion with individuals was used in places where density was very low. This is a further illustration of the way in which the one tendency that consistently distinguished between patterns of consultation with office workers compared with manual workers was the greater disposition of managers to talk with individual office workers.

Consultation over office changes in relation to the content of the change

When we looked at consultations over office change in relation to its reported impact upon staffing, earnings and job interest, we found that the impact of the change upon staffing was the aspect most strongly associated with levels and forms of consultation (see Table VI.17). There was a general tendency for different forms of consultation to be more common in circumstances where staffing was reduced as a result of the change, but the tendency was especially marked in relation to discussions with union officers. So far as both shop stewards and full-time officers were concerned, discussions were most common where manning was reduced, less common where there was no change, and least common where manning was increased.

Table VI.15 Extent and form of consultations over the introduction of advanced technical change in the office, in relation to union recognition

Column percentages

	All establishments		Establishments with fewer than 50 non-manual workers	
	Non-manual union recognised	Union not recognised	Non-manual union recognised	Union not recognised
One or more of listed forms of consultation	94	83	91	75
No consultations	4	15	5	22
Not stated	2	2	4	3
Formal—union channels[a]				
Discussed with shop stewards	35	–	23	–
Discussed with full-time officers	16	–	14	–
Formal—non-union channels				
Discussed in established JCC	17	6	8	3
Discussed in specially constituted committee	16	12	18	10
Informal				
Discussions with individual workers	73	79	66	74
Meetings with groups of workers	46	43	46	41
Base: establishments with 25 or more non-manual workers that introduced advanced technical change in the office				
Unweighted	*747*	*230*	*55*	*74*
Weighted	*286*	*213*	*83*	*119*

[a] See note F.

It was not really possible for us to explore the extent to which any impact which the change had upon earnings was associated with levels of consultation, because reductions in pay resulting from office change were so infrequently reported. So far as the level of the decision was concerned, it was clear that formal consultations were much more common

Table VI.16 Extent and form of consultations over the introduction of advanced technical change in the office, in relation to union density

Column percentages[a]

	Non-manual union density				
	1-24	25-49	50-89	90-99	100
One or more of listed forms of consultation	(89)[b]	98	96	98	96
No consultations	(11)	2	3	2	3
Not stated	–	–	1	–	1
Formal—union channels[c]					
Discussed with shop stewards	(7)	22	43	33	55
Discussed with full-time officers	(4)	15	11	22	35
Formal—non-union channels					
Discussed in established JCC	(5)	19	14	25	17
Discussed in specially constituted committee	(6)	26	11	16	15
Informal					
Discussions with individual workers	(83)	61	76	84	67
Meetings with groups of workers	(31)	50	46	55	51
Base: establishments with 25 or more manual workers that introduced advanced technical change in the office					
Unweighted	*47*	*87*	*275*	*77*	*80*
Weighted	*22*	*39*	*97*	*34*	*34*

[a] See note C.
[b] See note B.
[c] See note F.

when the change was initiated at a level higher than the establishment. This was especially the case with regard to discussions with union officers (see Table VI.17) and it remained true when analysis was confined to places that recognised trade unions.

The extent of negotiations over technological change affecting office workers
Overall, advanced technical change affecting office workers was less likely to be negotiated with trade union representatives than change

Table VI.17 Extent of consultation with union officers over introduction of advanced technical change in the office, in relation to some characteristics of the change

Column percentages

	Impact on staffing			Level of decision to introduce change			
	Increased	No change	Decreased	Independent establishment	Local decision	Joint decision	Central decision
One or more of listed forms of consultation	87	91	92	86	91	100	90
No consultation	13	9	8	14	9	*	10
Not stated	–	*	*	4	*	–	*
Formal consultations with union							
Discussed with shop stewards	12	19	32	4	17	30	31
Discussed with full-time officers	4	9	12	2	7	16	16
Base: establishments with 25 or more manual workers that introduced advanced technical change in the office							
Unweighted	*79*	*638*	*245*	*72*	*502*	*75*	*297*
Weighted	*50*	*351*	*88*	*79*	*222*	*37*	*140*

affecting manual workers, chiefly because non-manual unions were less likely to be recognised. When analysis was confined to workplaces where the respective unions were both recognised, then the extent of joint regulation of change was remarkably similar (see Table VI.18). The overall pattern of variation in the extent to which technological change in offices was subject to a joint regulation and joint consultation with trade union representatives closely followed variations in the pattern of recognition for non-manual unions. The variations of interest, then, lay in the differences that persisted among establishments where non-manual unions were recognised. On this basis, there was no simple association between the extent of negotiation and size, either the size of establishments in terms of the number of non-manual workers employed or the size of organisations measured by the total number of employees. Negotiations over change tended to be most common in circumstances where intermediate numbers of non-manual workers were employed (see Table VI.19).

Negotiations tended to be more common in the public than in the private sector (see Table VI.20). They were especially common in nationalised industries. In the private sector, remarkably, negotiations were reported more frequently in the service sector than in manufacturing, when analysis was confined to places where non-manual unions were recognised. Retail distribution and transport contributed particularly strongly to the pattern. There was no consistent pattern in relation to nationality of ownership or whether or not an establishment was part of a larger organisation.

There was a strong tendency for negotiations to be more common the higher the level of trade union density, especially when places of similar size were compared (see Table VI.21). In terms of the form and content of the change there were tendencies for negotiations to be more common where pay was increased as a result of the change, where manning was decreased and where the decision to introduce the change was taken at a higher level than the establishment.

Non-manual stewards' accounts of forms and levels of consultation and negotiation over the introduction of advanced technical change

The relatively favourable picture of consultations with office workers over the introduction of advanced technical change according to the accounts of managers was made slightly less favourable by the addition of the reports of non-manual stewards as shown in Table VI.22. The favourable picture rested largely on the greater extent to which managers reported informal consultations with office workers than with manual workers, and particularly the greater extent to which they reported discussions with individual workers. When we spoke with non-manual stewards in the minority of cases of technical change in offices where the

Table VI.18 Extent of negotiations over advanced technical changes affecting manual workers compared with changes affecting office workers

Column percentages[a]

	All relevant establishments		Establishments where advanced technical change affected both categories of employee	
	Manual workers	Office workers	Manual workers	Office workers
Relations with union representatives over change				
(A) In all relevant establishments				
Change negotiated	6	5	9	6
Consultations with union representatives over change	46	25	55	32
No discussion	23	21	16	23
Not stated/did not apply	25	49	20	39
(B) In establishments where unions were recognised				
Change negotiated	8	9	11	10
Consultations with union representatives over change	61	44	68	54
No discussion	30	39	20	34
Not stated/did not apply	1	8	*	2
Base: as column heads[b]				
Unweighted	*405*	*736*	*269*	*269*
Weighted	*160*	*283*	*48*	*48*

[a] See note C.

Table VI.19 Extent of negotiations over advanced technical change in the office in relation to the number of non-manual workers employed

Column percentages

	All establishments	Number of non-manual workers at establishment 25-49	50-99	100-199	200-499	500-999	1000 or more
Base A: all establishments with non-manual workers							
Change negotiated with union representatives	5	*	7	9	13	8	6
Consultation with union representatives	25	17	23	31	38	50	55
Unweighted	*977*	*129*	*165*	*220*	*223*	*137*	*103*
Weighted	*500*	*201*	*137*	*91*	*47*	*14*	*10*
Base B: non-manual union recognised							
Change negotiated with union representatives	9	1	12	13	17	10	6
Consultation with union representatives	44	44	39	44	49	60	58
Unweighted	*747*	*55*	*114*	*171*	*192*	*119*	*96*
Weighted	*288*	*83*	*83*	*64*	*37*	*11*	*10*

Table VI.20 Extent of negotiations over advanced technical change in the office in relation to sector

Column percentages

	Private manu-facturing	Private services	Nation-alised industries	Public services
Base A: establishments with 25 or more non-manual workers experiencing advanced technical change in the office				
Change negotiated with union representatives	3	3	21	9
Consultation with union representatives	19	13	72	45
Unweighted	*348*	*255*	*105*	*269*
Weighted	*121*	*231*	*24*	*124*
Base B: establishments that recognised non-manual unions				
Change negotiated with union representatives	6	8	21	9
Consultation with union representatives	41	36	72	45
Unweighted	*257*	*116*	*105*	*258*
Weighted	*55*	*84*	*24*	*121*

views of shop stewards were relevant, they were substantially less likely than managers to report informal consultations with individuals and groups of workers. In consequence, the proportion of cases where there was no consultation was very much larger according to stewards than it was according to managers, in the same workplaces. Surprisingly, non-manual stewards were more likely than managers to report consultation through an established joint consultative committee. This was the only point at which stewards, manual or non-manual, reported more of any type of consultation than did managers. When analysing the answers of non-manual stewards in this section and in the next on the extent of negotiations, we had a slight sense that there might have been a tendency among some of them to seek to establish themselves as employee representatives and hence to minimise the role of informal, personal consultations and emphasise the role of more formal channels. Equally, however, it is likely that there was a similar tendency on the part of managers to make the most of any discussions with individuals which they had carried out.

As was the pattern for manual workers, the differentiation in our questions asked of stewards between consultation at different stages further

Table VI.21 Extent of negotiations over advanced technical change in the office in relation to union density

Column percentages[a]

	Establishments with 50–149 non-manual workers			Establishments with 150 or more non-manual workers		
	Non-manual union density			Non-manual union density		
	1-49 per cent	50-99 per cent	100 per cent	1-49 per cent	50-99 per cent	100 per cent
Change negotiated with union representatives	(8)[b]	15	(19)	2	17	(18)
Consultation with union representatives	(31)	48	(33)	52	51	(64)
No discussions with union representatives	(53)	37	(48)	46	28	(18)
Not stated	(8)	*	–	–	3	(*)
Base: establishments with 25 or more non-manual workers that recognised non-manual unions and introduced advanced technical change in the office						
Unweighted	*38*	*106*	*27*	*70*	*249*	*39*
Weighted	*20*	*61*	*10*	*13*	*40*	*9*

[a] See note C.
[b] See note B.

reduced any impression given by managers of comprehensive consultation. Table VI.23 summarises non-manual stewards accounts of consultations at each of the two stages separately. Table VI.24 shows the proportion of cases where particular types of consultation were reported at both stages. Whereas stewards reported that individuals were consulted at *some* stage in about one half of cases, they reported that they were consulted at *both* stages in only 22 per cent of cases. As was the case with the manual pattern, private services emerged as the sector least likely to have anything approaching comprehensive formal consultations over the introduction of technical change.

Non-manual stewards' reports of negotiations over change

The answers of stewards to our question seeking to establish how far the introduction of advanced technical change in the office was subject to joint regulation were rather paradoxical compared with those of managers in places where non-manual unions were recognised. According to

Table VI.22 Comparison of managers' and non-manual stewards' accounts of the extent and form of consultation over the introduction of advanced technical change in the office

Column percentages[a]

	Union recognised		Establishments where both were interviewed about an advanced technical change	
	All managers reporting advanced technical change	All stewards reporting advanced technical change	Managers	Stewards
One or more of listed forms of consultation	94	79	94	83
No consultation	4	19	2	16
Not stated	2	2	4	1
Formal — union channels[b]				
Discussed with shop stewards	35	34	49	36
Discussed with full-time officers	16	20	20	17
Formal — non-union channels				
Discussed in established JCC	17	30	23	32
Discussed in specially constituted committee	16	16	18	18
Informal				
Discussions with individual workers	73	50	71	50
Meetings with groups of workers	46	34	53	33
Base: as column heads[c]				
Unweighted	*747*	*520*	*437*	*437*
Weighted	*286*	*201*	*138*	*138*

[a] See note C.
[b] See note F.
[c] See note to Table VI.7.

stewards, union representatives were in some way involved in the change slightly less frequently than they were according to managers. But where they were involved, stewards were more likely to report that they were engaged in negotiations. The pattern was inconsistent with any assumption of a neat progression from no involvement through consultation to negotiations. In fact, it was likely to reflect underlying negotiating po-

Table VI.[illegible] Non-manual stewards' accounts of the extent and form of consultation over the introduction of advanced technical change in the office

Column percentages[a]

	Total		Private manufacturing		Private services		Nationalised industries		Public services	
	Before	During	Before	During	Before	During	Before	During	Before	During
One or more of listed forms of consultation	70	62	58	60	79	57	67	68	72	62
No consultations	28	33	41	39	21	35	27	19	27	33
Not stated	2	6	*	1	–	8	6	13	2	4
Formal — union channels[b]										
Discussed with shop stewards	26	25	25	18	8	19	36	38	31	26
Discussed with full-time officers	15	15	5	5	3	12	32	24	19	16
Formal — non-union channels										
Discussed in established JCC	25	23	12	12	15	25	41	24	29	26
Discussed in specially constituted committee	10	13	8	9	4	17	14	16	12	13
Informal										
Discussions with individual workers	42	30	39	39	59	27	29	31	40	29
Meetings with groups of workers	28	23	24	21	19	17	23	18	33	27
Base: non-manual stewards who reported advanced technical change										
Unweighted	*520*	*520*	*169*	*169*	*66*	*66*	*78*	*78*	*207*	*207*
Weighted	*201*	*201*	*33*	*33*	*36*	*36*	*25*	*25*	*107*	*107*

[a] See note C.
[b] See note F.

Table VI.24 Non-manual stewards' accounts of consultations both before the introduction of advanced technical change and during its implementation

Column percentages

	Total	Private manufacturing	Private services	Nationalised industries	Public services
Formal — union channels					
Discussed with shop stewards	17	15	7	26	19
Discussed with full-time officers	10	4	1	20	12
Formal — non-union channels					
Discussed in established JCC	18	9	14	21	22
Discussed in specially constituted committee	8	5	4	8	10
Informal					
Discussions with individual workers	22	22	20	20	23
Meetings with groups of workers	17	17	6	12	21
Base: non-manual stewards who reported advanced technical change					
Unweighted	*520*	*169*	*66*	*78*	*207*
Weighted	*201*	*33*	*36*	*25*	*107*

sitions. Traditionally, trade unions have been concerned to push back the frontiers of joint regulation by increasing the agenda of collective bargaining. Consequently, when topics have come up for discussion with managements, they have sought to make them the subjects of negotiation. One of the tactics used in seeking to extend the boundaries of negotiation has been to talk the language of negotiations. In contrast, managers have traditionally sought to preserve the managerial prerogative to decide. When in discussion with union officers, they have consequently been disposed to operate on the basis that they were consulting worker representatives and would subsequently decide on the basis of those consultations together with all the other relevant considerations. Accordingly, the pattern in Table VI.25 may well be a reflection of the tactical inclination on the part of union representatives to talk the language of negotiation

Table VI.25 Comparison of managers' and non-manual stewards' accounts of the extent of negotiations over advanced technical change in the office

Column percentages[a]

	Non-manual union recognised		Workplaces where both managers and stewards were interviewed about an advanced technical change	
	All managers	All shop stewards	Managers	Stewards
Change negotiated with union representatives	9	16	12	16
Consultation with union representatives	44	48	55	48
No discussions with union representatives	39	34	31	34
Not stated	8	2	1	2
Base: as column heads[b]				
Unweighted	*747*	*520*	*437*	*437*
Weighted	*286*	*201*	*138*	*138*

[a] See note C.
[b] See note to Table VI.7.

and the countervailing inclination on the part of managers to talk the language of consultation. This explanation is certainly plausible and fits the pattern of modest contrast in the accounts of managers and non-manual stewards of the process in which they were engaged. Unfortunately, as we saw earlier, when analysing levels of consultation and negotiating involved in advanced technological change affecting manual workers, the separate accounts of manual stewards and managers did not reveal a similar pattern to that in Table VI.25. This was certainly true when analysis was confined to cases where we had the reports of both. It may well be, however, that relations between managements and manual unions were generally longer established and more stable than those between managements and non-manual unions. The implication would be that it was when unions were seeking to establish themselves and the bargaining agenda, and when managements were acting to preserve their prerogatives, that each had an inclination to talk as if the world was as it wanted it to be.

Footnotes

(1) The most recent major public debate about worker involvement in decisions arose over the Bullock report (*Report of the Committee of Inquiry on Industrial Democracy*, HMSO, 1977). All parties to the debate agreed that worker involvement was desirable. The focus of dispute was the level and means of involvement. See John Elliott, *The Growth of Industrial Democracy*, Kegan Page, 1978.

(2) Paul Blumberg, *Industrial Democracy: The Sociology of Participation*, Constable, London, 1968.

(3) F. Roethlisberger and W.J. Dixon, *Management and the Worker*, Wiley, New York, 1964; L. Coch and J.R.P. French, 'Overcoming Resistance to Change', *Human Relations*, No.1, 1948.

(4) Shirley Williams, *Politics is for People*, Penguin Books, Harmondsworth, 1981.

(5) Paul Willman, *New Technology and Industrial Relations* — a Review of the Literature, Department of Employment, Research Paper, No. 56, 1986.

(6) A good example is provided by the measures adopted by ICI to introduce their Weekly Staff Agreement (WSA). One case study of that agreement was provided by S. Cotgrove, J. Dunham and C. Vamplew, *The Nylon Spinners*, Allen and Unwin, London, 1971.

(7) TUC, 1981, *op.cit.*

(8) W.W. Daniel and Neil NcIntosh, *The Right to Manage*, Macdonald, London, 1972.

(9) W.W. Daniel and Neil Millward, *Workplace Industrial Relations in Britain*, Heinemann, London, 1983.

(10) N.E.J. McCarthy and N.D. Ellis, *Management by Agreement*, Hutchinson, London, 1973.

VII Technical Change and the Quality and Organisation of Work

Thinking about industrial work in the first two thirds of the twentieth century was dominated by the mass production assembly line. Blessed by the ideas of Adam Smith(1) and nurtured by those of Frederick Taylor(2), the assembly line came to symbolise the dehumanisation of work in both popular entertainment and serious social commentary. The motor car assembly line came to epitomise all that was worst in that system of production(3). Workers were required to spend their working day doing one simple repetitive task, devoid of interest, meaning and satisfaction. They were physically isolated from their fellow workers and denied social contact and companionship. They were slaves to the line and while it ran they could not stop. The psychological and social deprivations exacted by the assembly line were often accompanied by physical discomfort as a consequence of heat, dirt, noise and the requirement to spend long periods in stressful positions.

For some commentators the advent or prospect of automation promised liberation from the assembly line. At the simplest level, they argued, it was better for mindless tasks to be done by creations without minds like sophisticated electronic equipment or robots. At a more sophisticated level, some argued that work on new automated processes represented the recreation of work that had meaning, interest, satisfaction and dignity. In particular, Robert Blauner promoted the view that automation heralded an end to the meaninglessness, isolation, powerlessness and self-estrangement of work on the assembly line(4). Automated processes were built on the principle of integration, while the assembly line was derived from the concept of fragmentation. Operators of automated processes took a pride in their responsibility for large, complex, modern plant. They could move around freely, socialise and take breaks when they wished. Automated workshops were relatively clean, quiet, light, airy, providing a moderate temperature and much improved physical working conditions compared with, for instance, the motor car assembly line. Blauner took the process production of heavy chemicals or oil refining as his model for the automated factory that would replace the traditional mass production workshop. He showed how many of the benefits he attributed to the system of production had been experienced by chemical workers in the United States. Support for his conclusions also came from interviewing with similar workers in Britain(5).

Not all commentators shared Blauner's rosy view of the likely impact of automation upon work. Some argued that, on the contrary, it took the process of dehumanising work a stage further than the assembly line(6). Automation meant that man was even more dominated by the machine. The process was in control and man was required to react in an unthinking way without any real knowledge or understanding of the process. Automation required human robots. Particular horrors were attributed to automation in the office as compared with automation on the shop floor(7). In fact, despite great controversy about the implications of automation for the quality of work, very little systematic research has been carried out on how workers experience the implications of processes that include microelectronics. Case studies carried out by the Work Research Unit, however, suggest, first, that their application can reduce certain types of skill and sources of interest in work, but that they also provide opportunities for others(8).

Managements, however, all too often tended to focus upon the technical questions raised by the introduction of change and to ignore the human implications of choices. This conclusion certainly echoed themes in our own analysis. As we showed in earlier chapters, personnel departments were rarely involved at an early stage of the introduction of technical change and levels of worker involvement in change were low. At this stage we are able to provide some evidence of the general implications of the new technology for the content of work. The strength of this evidence is that, for the first time, a picture is provided of the impact of microelectronics upon intrinsic rewards from work that covers the economy as a whole. The limitations of the analysis will become apparent as we go along. We start with the accounts by managers and shop stewards of how the most recent advanced technical changes at their workplaces affected the job content of manual workers. We continue in the second half of the chapter with the parallel accounts of the impact of computers and word processors upon the jobs of office workers.

Advanced technology and the job content of manual workers

A major limitation of our information on the effects of advanced technology on the content of the jobs done by the workers affected is that we did not collect any information from the workers themselves. Accordingly, we cannot pretend to be reporting the experiences and feelings of workers actually engaged upon the new machines or plant. On the other hand, we do have the reports of shop stewards, of our principal management respondents, who tended to be personnel managers in larger workplaces and general managers in smaller places, and also, so far as the majority of manual applications were concerned, of works managers. In most cases, each of these three types of respondent will have been involved in directly managing or representing workers engaged in advanced technology. So far as stewards were concerned, it was almost certainly the case that some of them would themselves have worked with the new technology. If we found consistency in the separate, independent accounts of the impact of the changes, from quite distinct perspectives, this would provide persuasive evidence of their general effects upon jobs. Our main questioning in this area consisted of asking both managers and shop stewards a battery of questions seeking to establish whether the change had increased, decreased or had no effect upon: level of skill; range of activity; level of supervision; job interest; level of responsibility; pace of work; and workers' influence over how they did their jobs. So far as works managers were concerned, in manufacturing industry we supplemented these questions with further enquiries about changes that had been taking place in the organisation of work at their plant. In particular, we were interested in any measures taken to promote flexibility of labour and how far any such measures were associated with new technology. Following our report of managers' and stewards' accounts of the impact of advanced technology upon the jobs of both manual workers and office workers, we conclude the chapter with a more detailed analysis of changes in work organisation in manufacturing industry.

We have already indicated in Chapter IV that, according to the reports of managers, advanced technological change appeared to have a generally favourable impact upon the content of both manual and office jobs. The impact was substantially more favourable than the implications of the generality of changes in organisation that were taking place independently of technical change, such as productivity agreements. But it was slightly less favourable than the impact attributed to conventional technical change. Why that should have been so was not immediately clear. In principle, it might be expected that new advanced technology would change the nature of jobs more profoundly than new plant or equipment based upon new conventional technology. Overall, however, it appeared that new conventional technology was slightly more popular than new advanced technology, in general, we suggested, probably because of its greater familiarity. It was possible that the slightly greater job enrichment attributed to conventional technology arose from a halo effect surrounding that which had the combined attractions of being both new and at the same time familiar.

In this chapter we report more fully upon different accounts of the changes and other sources of variation in the impact of advanced technical change on job content. Table VII.1 shows the answers in full of principal management respondents to our battery of questions in relation

Table VII.1 Managers' accounts of the impact of advanced technical change upon the jobs of manual workers

Column percentages

	Job interest	Skill	Range of activity	Respon-sibility	Pace of work	How they do their jobs	Super-vision
More	46	42	38	33	22	21	16
No change	40	42	46	54	48	47	64
Less	11	15	15	12	28	31	19
Not stated	3	1	1	1	2	1	1
Change score[a]	+70	+50	+50	+40	−10	−20	+10[b]
Base: establishments with 25 or more manual workers and experiencing advanced technical change							
Unweighted	*458*	*458*	*458*	*458*	*458*	*458*	*458*
Weighted	*212*	*212*	*212*	*212*	*212*	*212*	*212*

[a] See note I.
[b] A positive score on this item represents *less* supervision.

to the jobs of manual workers affected by advanced technical change. It is clear, first, that in many instances managers reported that the technical change had no effect on any of the nominated aspects of jobs. We need to emphasise this feature because there is a danger that it may be lost in the system of abbreviated scores that we subsequently adopt for purposes of exploring the pattern of variation as between different categories of workplace. In most instances, managers reported no change in relation to nearly one half of the items. The second feature of Table VII.1 is that there was a group of items which clearly emerged as having been enhanced by the change according to the managers' accounts. These related to job interest, skill, range of activity or task repertoire and level of responsibility. In each instance, many more managers reported a favourable impact than an unfavourable one. In consequence, there was a strongly positive change score associated with each.

Thirdly, there emerged a group of items where, according to the reports of managers, the impact of the change was more neutral and, indeed, where on balance it tended to be very slightly negative. Essentially, this group included the items concerned with worker autonomy: the control they had over the pace of work and over the way they did the job, and the level of supervision to which they were subject. It is clear that the notion of control is a complex one. This may be illustrated by taking the difference between mass production and automation according to

Blauner's analysis which we summarised at the start of the chapter. Under automation, certainly in the form taken by the automation of heavy chemical production, the worker has more freedom than under mass production, to move around, to talk to other people, to schedule the different activities to be done over the shift, including breaks, and indeed, to pay little or no attention to what is happening to the production process for short periods. In these senses the worker has more control over how he spends his time while at work and over the speed at which he works. At the same time, because under automation the process is less dependent upon human activity and intervention than under mass production, management also has more control over the execution of the process and its speed. When, in our results, managers said that workers had less control over the way they did their jobs and over their pace of work, it is not clear whether they were talking about a managerial enhancement of control over the process and its speed with a consequent loss of freedom for the worker, or whether they were describing an increase in managerial control over the process and its pace which at the same time left the worker with more freedom, as under automated chemical production. We cannot choose between these fine distinctions on the basis of our present information. It remains the case that, in managers' judgements, the changes tended, on balance, to be unfavourable to workers' control over the content and pace of their work. It would be dangerous to try and argue away this result, especially as it was consistent with the judgement of shop stewards, who in some cases would almost certainly have been employed upon the jobs that were subject to the change.

Moreover, it is especially instructive that loss of control over the content of jobs and, to a slightly lesser extent, loss of control over the pace of work were the only two aspects of jobs that were strongly and consistently associated with the size of workplaces (see Table VII.2). It is plausible that larger workplaces would be more inclined to use the new technology to increase managerial control over working tasks and the speed at which they were done and to reduce the control of individual workers. The strength and consistency of the pattern does also suggest that the impact of the new technology on jobs was not wholly predetermined. It implies some choice in how the technology was applied, and suggests that the larger the workplace the more likely was management to use its discretion to increase its control and reduce that of individual workers. An alternative explanation is that applications of the new technology had different implications for the control of workers over the content and pace of their work, depending upon the process to which they were applied, and that the applications to processes characteristic of larger workplaces were different from those characteristic of smaller workplaces.

Certainly, Table VII.3 shows that there were also major differences on

Table VII.2 Managers' accounts of the impact of advanced technical change upon the content of manual workers' jobs in relation to number employed

Scores and percentages

		Number of manual workers at establishment					
	Total	25-49	50-99	100-199	200-499	500-999	1000 or more
Job interest							
Mean score[a]	+70	+60	+80	+100	+60	+60	+80
Per cent more	*46*	*38*	*53*	*58*	*44*	*43*	*49*
Skill							
Mean score	+50	+60	+50	+50	+50	+30	+60
Per cent more	*42*	*44*	*41*	*45*	*37*	*35*	*44*
Range of tasks							
Mean score	+50	+60	+30	+60	+40	+50	+70
Per cent more	*38*	*37*	*32*	*49*	*39*	*45*	*42*
Responsibility							
Mean score	+40	+20	+50	+70	+60	+70	+60
Per cent more	*33*	*20*	*40*	*48*	*38*	*46*	*36*
How they do their job							
Mean score	−20	−10	−20	−30	−40	−70	−60
Per cent more	*21*	*19*	*25*	*21*	*19*	*18*	*12*
Pace of work							
Mean score	−10	−10	+10	−10	−40	−50	−50
Per cent more	*22*	*22*	*27*	*23*	*15*	*22*	*11*
Supervision							
Mean score[b]	+10	*	+10	+20	+20	+10	+50
Per cent more	*19*	*15*	*20*	*25*	*24*	*20*	*32*
Base: establishments with 25 or more manual workers and experiencing advanced technical change							
Unweighted	*458*	*50*	*56*	*70*	*121*	*104*	*57*
Weighted	*212*	*91*	*51*	*31*	*27*	*8*	*4*

[a] See note I.

[b] See note to Table VII.1.

Table VII.3 Managers' accounts of the impact of advanced technical change upon the content of manual workers' jobs, in relation to sector

Scores and percentages

	Total	Private manuf-acturing	Private services	Nation-alised industries	Public services
Job interest[a]					
Mean score	+70	+90	+50	+70	+20
Per cent more	*46*	*55*	*36*	*40*	*19*
Skill					
Mean score	+50	+40	+110	+80	+30
Per cent more	*42*	*37*	*57*	*46*	*37*
Range of tasks					
Mean score	+50	+50	+50	+80	−20
Per cent more	*38*	*45*	*30*	*42*	*16*
Responsibility					
Mean score	+40	+50	+30	+70	+10
Per cent more	*33*	*38*	*17*	*40*	*27*
How they do their job					
Mean score	−20	−50	+50	−70	+10
Per cent more	*21*	*14*	*36*	*8*	*35*
Pace of work					
Mean score	−10	−30	+20	−40	+30
Per cent more	*22*	*20*	*29*	*16*	*27*
Supervision[b]					
Mean score	+10	+30	*	−10	−50
Per cent less	*19*	*24*	*18*	*7*	*11*
Base: establishments with 25 or more manual workers and experiencing advanced technical change					
Unweighted	*458*	*286*	*59*	*63*	*50*
Weighted	*212*	*116*	*44*	*26*	*28*

[a] See note I.
[b] See note to Table VII.1.
* Neutral.

the items concerning control in relation to sector, according to the judgements of managers. In private manufacturing, it appeared that the impact upon items concerned with control was largely unfavourable to workers, while in private services it appeared generally favourable. In the public sector, the effects were much less favourable in nationalised industries than in public services. A substantial part of these variations, however, was accounted for by differences in the characteristic size of workplaces in the different sectors. For instance, when we isolated large workplaces, where there were sufficient numbers in private manufacturing, private services and public services for a satisfactory comparison to be made, the general impact upon control was similarly unfavourable in all three cases. There remained, however, a marked contrast between private manufacturing and service establishments of intermediate size.

While managers' accounts of the impact of advanced technology upon workers' control over the content and pace of work tended, on balance, to be slightly unfavourable, these were the only two items for which this was true. As we emphasised at the start of this section the judgements of managers generally revealed that the overall impact of the change was favourable. The reported effects of the change on job interest, skill, task repertoire and responsibility were substantially more favourable than judgements relating to control were unfavourable, and they were as favourable at large establishments as they were overall.

Reported impact upon job content in relation to union organisation

So far as variations in relation to union organisation were concerned, the only sources of systematic variation and the chief items of interest related to skill and responsibility in relation to union density. There was a marked and consistent tendency for managers to report lower increases in both skill and responsibility the higher the level of union density at workplaces. It may well be that this association owed much to the weight given to skill and responsibility in job evaluation or grading schemes, and the greater likelihood of claims for regrading having been pressed where union density was greater. Certainly, as we now go on to show, as we contrast management evaluations with those of shop stewards, there was a much greater tendency for union representatives to report increases in skill and responsibility than for managers to do so.

Stewards' accounts of the impact of advanced technical change upon job content

The addition of the perspective of the workers' union representatives at the workplace added three invaluable elements to our analysis of the impact of advanced technical change upon job content. First, as we emphasised earlier in the chapter, we were able to see whether there was consistency between the reports from quite distinct perspectives. If there was consistency, we could place more reliance upon the accounts given. Secondly, as also mentioned earlier, it is likely that at least some of the stewards themselves had worked on the plant or machinery whose impact they were describing or upon very similar plant. Thirdly, contrasts between managers' and stewards' accounts would give us some insight into issues that were likely to have been involved in discussions or negotiations between the parties over the change.

In the event, as Table VII.4 shows, the separate accounts of managers and stewards were sufficiently consistent to be reassuring but also sufficiently different to be interesting. As is our general practice in such

Table VII.4 Comparison of manual stewards' and managers' accounts of the impact of advanced technical change upon the job content of manual workers

Scores[a] and percentages

	Manual union recognised		Workplaces where both managers and stewards were interviewed about an advanced technical change	
	All managers	All shop stewards	Managers	Stewards
Job interest				
Mean score	+80	+60	+90	+60
Per cent more	*49*	*46*	*55*	*44*
Skill				
Mean score	+60	+110	+90	+120
Per cent more	*44*	*62*	*58*	*64*
Range of tasks				
Mean score	+50	+110	+80	+100
Per cent more	*37*	*64*	*48*	*64*
Responsibility				
Mean score	+40	+110	+50	+120
Per cent more	*31*	*58*	*36*	*63*
How they do their job				
Mean score	−20	*	−30	−10
Per cent more	*24*	*30*	*22*	*28*
Pace of work				
Mean score	*	−10	−10	−10
Per cent more	*25*	*26*	*29*	*26*
Supervision				
Mean score	*	+20	+10	*
Per cent less	*17*	*28*	*20*	*24*
Base: as column heads[b]				
Unweighted	*405*	*288*	*205*	*205*
Weighted	*160*	*101*	*72*	*72*

[a] See note I.
[b] See note to Table VI.7.
* Neutral

tables, we compare, in the first two columns, the reports of all relevant managers in places where unions were recognised with the reports of all relevant stewards that we interviewed. In the second two columns we isolate more strictly comparable cases where both stewards and managers were interviewed and both provided accounts of the impact of the introduction of advanced technical change upon the jobs of some group of manual workers at their workplace. It is clear from Table VII.4 that, whichever comparison was made, the overall impression created by the different accounts was broadly similar. According to both managers and stewards, the change resulted in substantial increases in job interest, skill, range of tasks and responsibility. Stewards also appeared to agree with managers that the impact of advanced technical changes was slightly negative in relation to their impact upon autonomy, as measured by level of supervision, and control over both the pace of working and how the job was done.

While there was broad agreement along these lines, there were also clear differences in detail that, as already suggested, may well have reflected underlying negotiating positions. For instance, stewards were substantially more inclined than managers to report that the change resulted in increases in skill, range of tasks and, especially, responsibility. But stewards were less likely than managers to judge that the change had increased the interest in the job. In contrast, managers were clear that the change enhanced intrinsic job interest but they were less inclined than stewards to feel that it increased levels of skill and responsibility and range of tasks. The most plausible explanation of these contrasts is that they result from negotiating positions regarding job grading. We report in Chapter X that a substantial minority of changes resulted in increases in earnings for the workers involved as a result of upgrading. It is very likely that requests or claims for regrading would have been put forward in other cases. Most job evaluation systems or grading schemes for manual workers include weightings for responsibility, skill and range of tasks. Consequently, the rank order of advantages attributed to the change by managers may have reflected a tendency on the part of some to emphasise the benefit derived by workers from any increase in intrinsic job interest and to moderate their accounts of the benefits in terms of skill, responsibility and range of tasks that might have given grounds for regrading. For their part, some stewards may have been inclined to minimise any rewards that workers might have derived from the change through an increase in intrinsic job satisfaction, and to emphasise any changes that might have provided grounds for upgrading. Both sets of accounts may be seen to contain slight inconsistencies. So far as stewards were concerned, the increases in responsibility, skill and range of tasks they reported were relatively so great that it is surprising that they reported relatively so little increase in intrinsic job interest, which is so heavily dependent upon such job ingredients(9). But equally, so far as managers were concerned, it is surprising that the increase in job interest they reported was relatively so great when their accounts of the increase in its constituents were relatively so modest.

A second point worthy of note in the contrast between the reports of managers and those of stewards relates to their respective accounts of how far the change permitted workers to have more control over the way they did their jobs. Managers were inclined to say that, on balance, there was a marginal reduction in such control for the workers. Stewards were less inclined to report such loss of control, when, if it had occurred, they might have been expected to lay more emphasis upon it. The pattern highlights the complexity of the idea of control over the way the job was done which we mentioned when first discussing the pattern of managers' answers to this battery. It may be that the reports of stewards were closer to the experience of the workers affected on this item. Overall, however, given the differences in the roles and perspectives of managers and stewards, and the overt or latent pressures there may have been upon each to represent the impact of advanced technological changes upon job content in particular ways, the most striking and remarkable feature of Table VII.4 is the similarity revealed between the reports of the two parties.

The impact of advanced technology upon the jobs of office workers

We recorded in Chapter III how the introduction of advanced technology has affected office workers more widely than manual workers. The accounts we received from managers and union representatives of the impact of the latest technical changes upon the job content of the workers affected suggested that such innovations also affected the jobs of office workers more profoundly. Table VII.5 shows in full managers' answers

Table VII.5 Managers' accounts of the impact of word processors and computers upon the jobs of office workers

Column percentages[a]

	Job interest	Skill	Range of activity	Responsibility	Pace of work	How they do their jobs	Supervision
More	60	55	59	39	34	32	10
No change	27	39	29	55	45	38	70
Less	5	2	8	2	16	26	17
Not stated	8	4	4	4	5	4	3
Change score[b]	+120	+110	+110	+80	+40	+10	+10[c]

Base: establishments with 25 or more non-manual workers and experiencing advanced technical change in the office

Unweighted	*977*	*977*	*977*	*977*	*977*	*977*	*977*
Weighted	*500*	*500*	*500*	*500*	*500*	*500*	*500*

[a] See note C.
[b] See note I.
[c] See note to Table VII.1.

to the questions about the effects of word processor and computer applications upon the job content of office workers. The corresponding table for manual workers was reproduced earlier (Table VII.1). Table VII.6 summarises the differences between manual and office workers revealed by our questions about changes in job content and also shows the contrast in workplaces where both manual and office workers experienced advanced technical change.

In combination, the tables show, first, that managers were substantially less likely to report that the technical change had no impact upon many aspects of the jobs of office workers. This was certainly the case so far as job interest, skill and task repertoire were concerned. Secondly, they were substantially more likely to report that the change had a favourable impact upon different aspects of the jobs of office workers than upon the jobs of manual workers. The contrast was especially marked in relation to range of tasks; intrinsic job interest; skill level; autonomy; and control over the pace of work. For instance, 59 per cent of managers reported that the range of tasks carried out by office workers was increased as a result of the change, compared with 38 per cent who reported that their most recent technical innovation on the shop floor had a similar impact upon the manual workers affected. Again, 60 per cent reported that the intrinsic job interest of office workers was enhanced as a result of the change, compared with 46 per cent for manual workers. As well as reporting more frequently that changes had a favourable impact upon the content of office jobs, managers reported less frequently that they had an unfavourable impact. In consequence, the average score relating to each item and shown in the third section of Table VII.5, was generally positive in the case of office workers. This was true of all dimensions of job content at which we looked, except for the level of supervision. On that item the scores were the same for both office and manual workers.

The second two columns in Table VII.6 focus upon those workplaces that had recently introduced advanced technical changes affecting both office and manual workers; that is to say, word processors or computers affecting office workers and separate microelectronic applications affecting manual workers. We found that the contrast in assessments of the impact of change upon the two categories was even more pronounced. As we have noted throughout in our direct comparisons between office and

Table VII.6 Impact of advanced technical change upon the jobs of manual workers compared with office workers

Scores and percentages

	All relevant establishments		Establishments where advanced technical change affected both categories of employee	
	Manual workers	Office workers	Manual workers	Office workers
Favourable change (percentages)				
Job interest	46	60	47	74
Skill	42	55	36	61
Range of tasks	38	59	38	67
Responsibility	33	39	40	46
How they do their job	21	22	15	32
Pace of work	22	34	16	31
Supervision	19	17	24	21
Unfavourable change (percentages)				
Job interest	11	5	18	4
Skill	15	2	18	4
Range of tasks	15	8	20	10
Responsibility	12	2	12	2
How they do their job	31	26	46	36
Pace of work	28	16	41	15
Supervision	16	10	15	9
Average scores[a]				
Job interest	+70	+120	+60	+150
Skill	+50	+110	+30	+120
Range of tasks	+50	+110	+40	+120
Responsibility	+40	+80	+60	+90
How they do their job	+20	+10	−60	−10
Pace of work	+10	+40	−50	+30
Supervision	+10	+10	+20	+20
Base: as column heads[b]				
Unweighted	*458*	*977*	*326*	*326*
Weighted	*212*	*500*	*78*	*78*

[a] See note I.
[b] See note to Table V.5.

manual workers who worked in the same establishments, this contrast related to a particular sub-sample of workplaces, skewed towards the largest ones and concentrated in manufacturing industry. It was, however, in just that type of workplace that the bulk of manual workers were employed. Consequently, it appeared that the lower level of benefit being experienced by manual workers compared with office workers as a result of advanced technical change was most pronounced in places where manual workers were most numerous.

We have seen that according to managers' reports the introduction of advanced technology appeared to enrich the jobs of office workers substantially more than those of manual workers. When, however, we compared the reports of non-manual and manual stewards, we found that the contrast was not so great. There might be many explanations of this difference, but one contribution could be that our principal management respondents were quite simply closer to office workers physically and socially. We discussed this possibility in Chapter VI, where we reported that individual discussions and consultations between managers and workers over the introduction of changes were very much more common at the office level. Its relevance here is that managers' accounts of the comparative impact of advanced technology on the jobs of office and manual workers may have slightly over-stated the difference.

Variations in the impact of word processors and computers upon the job content of office workers

Our analysis of variations in managers' accounts of the impact of word processors and computers upon the jobs of office workers was rather unrewarding. There were few signs of clear or revealing patterns. Unlike the trend we identified in relation to manual workers, there were no indications of any consistent tendency for reported increases in autonomy to decline the larger the number of people employed. It did appear, however, that there was a less strong tendency for the task repertoire of office workers to increase in larger workplaces. There was some hint that the favourable effects of technical innovations upon jobs were more fully enjoyed in public services than in the private sector, especially compared with manufacturing. But none of these tendencies were very strong or consistent.

Perhaps the most interesting feature of the pattern was the interrelationships between the reported impact on the content of jobs and accounts of other features of change including the effects on earnings and staffing levels. Overall, increases in pay tended to be generally associated with favourable effects on the content of jobs. But the association was especially strong so far as increases in responsibility, skills and job repertoire were concerned. Given the nature of many job evaluation schemes, that is just what might have been predicted but the confirmation of such predictions provides further evidence for the robustness of our findings. So far as changes in staffing levels were concerned, the feature most strongly associated was task repertoire. This was more likely to have been increased in instances where staffing was reduced. It also ap-

Table VII.7 Comparison of non-manual stewards' and managers' accounts of the impact of advanced technical change upon the job content of office workers

Mean scores

	Non-manual union recognised		Workplaces where both managers and stewards were interviewed about an advanced technical changes	
	All managers	All shop stewards	Managers	Stewards
Job interest				
Mean score[a]	+120	+60	+130	+50
Per cent more	*63*	*47*	*64*	*43*
Skill				
Mean score	+100	+130	+110	+120
Per cent more	*52*	*70*	*54*	*65*
Range of tasks				
Mean score	+100	+80	+100	+80
Per cent more	*58*	*56*	*55*	*54*
Responsibility				
Mean score	+70	+90	+70	+90
Per cent more	*35*	*53*	*37*	*50*
How they do their job				
Mean score	+20	−30	+10	−30
Per cent more	*32*	*22*	*33*	*24*
Pace of work				
Mean score	+30	−10	+20	−20
Per cent more	*31*	*22*	*26*	*19*
Supervision				
Mean score	+10	−10	+10	*
Per cent less	*14*	*16*	*16*	*18*
Base: as column heads[b]				
Unweighted	*747*	*520*	*437*	*437*
Weighted	*286*	*201*	*138*	*138*

[a] See note I.
[b] See note to Table VI.7.
* Neutral.

peared that there was a strong positive association between increases in task repertoire and job enrichment as measured by intrinsic job interest.

Non-manual stewards' accounts of the impact of word processors and computers upon the content of office workers' jobs

Table VII.7 compares the accounts of non-manual union representatives of the implications of introducing word processors and computers upon the jobs of office workers with the reports of managers. We adopt our normal practice in such tables, of showing in the first two columns the overall pictures revealed by the two sets of independent accounts. In the second two columns, we compare union and management answers in places where both were interviewed about the impact of particular advanced technical changes. It is apparent, first, that there was a similar pattern to that revealed by the comparison of management and union accounts of the impact of change upon the jobs of manual workers, and which we attributed to underlying bargaining stances or preconceptions. Managers tended to emphasise the way in which the change enriched jobs through increasing intrinsic job interest, but to play down the ways in which levels of skill and responsibility and task repertoire were increased. Trade union representatives tended to play down the way in which office workers might have derived benefit through spontaneous job enrichment, and to emphasise those changes that tend to be associated with increases in pay, where a case can be made that they have occurred. On the other hand, the contrast between management and union accounts was not so great as it was in relation to manual jobs. This provided a further sign of greater closeness or affinity between our management respondents and office workers compared with their distance from manual workers. Secondly, there was a contrast between management and union respondents in relation to the implications of the change for autonomy, that was not apparent in our comparison at the manual level. Managers tended to report that the change marginally increased office worker control over the pace of work and over the way in which they did their jobs. Stewards, in contrast, tended to report a modest reduction in office worker autonomy.

Flexibility of working and the introduction of new technology in manufacturing industry

Since the early 1960s critics of Britain's poor industrial performance have identified wasteful working practices as one of the major causes. Such critics identified demarcations among different categories of craftsmen and between craftsmen and process workers as a cause of very substantial overmanning. The slogan 'half-time Britain on half pay' managed to appeal to both our national masochism and greed, at the same time(10). The diagnosis inspired the practice of productivity bargaining, the negotiation by management of changes in working practice contributing to more efficient work in conjunction with the negotiation of enhanced rates of pay which enabled workers to share in the increased productivity arising from the changes(11). The practice of productivity bargaining was given a royal seal of approval by the Donovan Commission and what very nearly amounted to an official kiss of death as a result of being actively encouraged by the government pay policy of the late 1960s(12). Nevertheless, despite being subject to the vagaries of fashion, the practice of productivity bargaining has continued for the past twenty years. More recently management spokesmen have argued that the needs of advanced technology in industry have made more urgent the breakdown of traditional demarcations between the tasks that may be done by different categories of worker and the promotion of greater flexibility of working among craft and process workers. Critics of managements have occasionally argued that they have used the introduction of advanced technology to bring about greater flexibility of working by the back door (13). The implications of such criticisms are that managements have outwitted workers by using the introduction of new technology to get them to adopt working practices that would not have been acceptable within the framework of conventional productivity bargaining, and certainly for a lower price than would have been required by productivity bargaining.

We set out to explore some of these issues in our present enquiry by looking, first, at how far managers felt there were constraints upon their freedom to organise work as they would like and why; secondly, how far managers had taken the initiative to try and promote a relaxation of a number of specified demarcations between different categories of manual work in the previous four years; and thirdly, we sought to establish the degree of flexibility in the existing working practices of process workers and craftsmen. We were then able to relate answers to these questions on flexibility of working to the experiences of workplaces in the introduction of advanced technology. Owing to the limits of space in our main management schedule, these detailed questions on work organisation had to be confined to our interviews with works managers. Consequently, our analysis of the implications of the introduction of advanced technology is confined to those parts of manufacturing industry that had both personnel managers and works managers. Essentially, these were the larger manufacturing workplaces. Nevertheless, although the limits to our information on these questions were a disadvantage, the larger establishments in manufacturing industry are of especial interest so far as the questions are concerned. Moreover, the answers were especially revealing. They showed that the introduction of advanced technology was indeed very strongly associated with more flexible patterns of working on the shop floor, such as those traditionally sought through productivity bargaining in unionised sectors of industry. Our results showed, further, that when increased flexibility was achieved through the introduction of advanced technology, the process was very much more acceptable to both workers and stewards than when it was achieved through productivity bargaining. Before examining these findings in some detail, however, it will be useful, as before, to look at the general distribution of the answers to our questions on flexibility and at the pattern of variation in answers in relation to the level of trade union organisation at the workplace. This will provide a context within which to locate our striking results on flexibility in relation to the use of new technology.

The general pattern regarding flexibility of working in manufacturing industry

Overall, about one third of our works managers felt that managers in their factories experienced constraints in their freedom to organise the distribution of tasks among different categories of process and maintenance work as they would have liked (see Table VII.8). This proportion is substantially lower than the 47 per cent of works managers who in the late 1960s thought that the organisation and arrangement of work in their plants could have been improved if they had been free to arrange their labour force as they wished, according to the Government Social Survey

Table VII.8 Summary[a] of works managers' answers to questions on flexibility of working

Percentages

		Number of manual workers at establishment			
	Total	Fewer than 200	200-499	500-999	1000 or more
Management constrained in organisation of work	34	26	46	57	(61)[b]
Flexibility of production workers on selected items					
Routine maintenance	28	31	21	21	(21)
Setting up machines	66	67	59	70	(70)
Flexibility of maintenance workers on certain items					
Working without mates	55	46	71	84	(70)
Mechanical fitting	55	58	53	44	(33)
Steps taken to promote flexibility	43	35	57	65	(70)
Relaxation of craft/craft demarcations	29	20	47	54	(51)
Base: all works managers					
Unweighted	*276*	*84*	*81*	*77*	*34*
Weighted	*76*	*52*	*16*	*6*	*2*

[a] These are items selected from among responses that are listed in full in Tables VII.10–VII.14.

[b] See note B.

report for Donovan (14). It is true that the question asked in that survey was different from our own, but the samples were similar and the fall over the period in the proportion reporting constraints was substantial. Moreover, in very nearly all instances where works managers reported any constraint in our present study, they also reported that they had sought to promote flexibility among manual workers in the recent period. Our results reported later in this section very strongly suggest that the introduction of advanced technology encouraged the relaxation of traditional demarcations. Accordingly, it would not be surprising if there had been some reduction in the limits to management discretion in the organisation of work over the past two decades in the face of the tide of technological innovation and management's commitment to reform.

As Table VII.8 also shows, there was a strong tendency for works managers to feel more constrained the larger the number of manual workers employed. Sixty-one per cent of works managers at places with 1,000 or more manual workers felt there were constraints on their freedom to organise work as they would like, compared with one quarter of counterparts at places employing fewer than 200 manual workers. This trend in managers' accounts of the extent to which they are constrained, however, was not always reflected in actual working practices. We also asked works managers whether production workers in their plants were normally involved, first, in the routine maintenance of machines or equipment, secondly, in the setting up of machines, thirdly, in the cleaning of machines or work areas, and, fourthly, in transporting materials to or from their work stations. Further, in an attempt to establish an operational measure of the flexibility of craftsmen, we asked whether electricians normally worked without semi-skilled mates, did mechanical fitting or maintenance, and worked on electronic equipment. So far as production workers were concerned, very nearly every works manager reported that they normally did at least one of the listed tasks. Cleaning, fetching and carrying were the activities most frequently reported. But whether they did routine maintenance and were involved in the setting up of machines served as more useful touchstones of the extent of operational flexibility in different types of workplace. As Table VII.8 shows, there was no marked tendency for production workers normally to operate more flexibly in those terms the smaller the size of the factory. There was more sign that the task repertoire of electricians was progressively more broad the fewer manual workers were employed. This was especially so in relation to the involvement of electricians in mechanical fitting; it was normal in more than one half of places employing fewer than 200 manual workers, but in only one third of the largest plants.

Our fourth main area of questioning concentrated upon whether managements had sought to increase the level of flexibility at their workplaces in the previous period. The particular types of possible initiative upon which we focussed our questions are listed in Table VII.13 below. Nearly one half of works managers reported that one or more of the listed initiatives to promote flexibility had been taken. Among these, whether or not managements sought the relaxation of demarcations as between different categories of craftsmen appeared to be the step which most clearly distinguished between managements that had been active in promoting flexibility and those that had not. It was apparent that managements in larger workplaces were more likely to report initiatives to make their workforces more flexible and they were especially likely to have sought some relaxation of demarcations as between groups of craftsmen. It is of particular interest in this analysis that demarcations between craftsmen should have been both the behavioural feature of work organisation that most effectively distinguished different types of workplace, and the area on which most management activity had concentrated.

Flexibility of working and trade union organisation

There was only limited scope for exploring issues concerning flexibility of working in relation to trade union organisation through our interviews with works managers. These interviews were necessarily confined to manufacturing industry and to manufacturing plants where there was a personnel function. Essentially these were the larger manufacturing units and most were characterised by a high level of trade union organisation. Of the 276 places where we interviewed works managers, only 29 did not recognise manual unions. In one half of the 276 places, all the manual workers were members of trade unions. There was little enough scope to look at the overall pattern in relation to union organisation and certainly not enough scope to examine the pattern within size bands.

Table VII.9 summarises the overall picture regarding our questions on flexibility of working and union organisation. Certainly, it appeared to be the case that managements felt substantially more constrained in their scope to organise work according to their own preferences in places where unions were recognised. Similarly, they felt more constrained the higher the trade union density in plants. Moreover, when we asked works managers who reported limits on their freedom whether they were constrained by any of the items summarised in Table VII.10, we found that union agreements, steward resistance and resistance from union members each ranked high in their lists. At the same time, it is noteworthy that lack of suitable skills among some sections of the workforce and the characteristic technology of the workplace also featured widely among constraints on management freedom to deploy labour as it saw fit.

Managers' accounts of the greater freedom they enjoyed to organise work where unions were not recognised were, however, not always reflected in the patterns of working among different work groups. For in-

Table VII.9 Summary[a] of works managers' answers to questions on flexibility of working in relation to union recognition and density

Percentages

	Manual union recognition		Manual union density		
	Union not recognised	Union recognised	50-89 per cent	90-99 per cent	100 per cent
Management constrained in organisation of work	(19)[b]	39	(32)	(39)	46
Flexibility of production workers on selected items					
Routine maintenance	(33)	26	(32)	(28)	25
Setting up machines	(49)	72	(71)	(66)	62
Flexibility of maintenance workers on selected items					
Working without mates	(23)	69	(78)	(60)	63
Mechanical fitting	(43)	60	(60)	(53)	53
Steps taken to promote flexibility	(20)	52	(62)	(63)	50
Relaxation of craft/craft demarcations	(2)	40	(53)	(48)	39
Base: all works managers					
Unweighted	*29*	*247*	*43*	*37*	*138*
Weighted	*22*	*54*	*11*	*6*	*28*

[a] See note to Table VII.8.
[b] See note B.

stance, it appeared that electricians were substantially more likely normally to work without mates and normally to do some mechanical work in trade union shops. Similarly, it appeared that the setting up of machines was more frequently a part of the normal duties of process workers in plants where unions were recognised. Managements in trade union workshops had certainly been substantially more active in the promotion of flexibility in recent years. This activity was especially marked so far as the relaxation of demarcations as between different categories of craftsmen was concerned. The general impression created by the picture in Table VII.9 is that managements in non-union workplaces were not inclined to take advantage of the freedom that was denied to counterparts in union shops to organise work as they would like, or certainly not inclined to take advantage of that freedom in the way that managers in trade union shops implied that they would.

The picture in relation to trade union density was slightly different, insofar as we were able to piece it together from not altogether satisfactory information. The cells in which we had sufficient numbers for analysis were confined to the top half of the continuum where more than one half of the workforce were trade union members. In that sector, however, which was generally highly organised, it did appear that the higher the density the greater the constraint on management's freedom to organise. Works managers reported more constraints and both process workers and electricians worked less flexibly the higher the level of union density. It also appeared to be the case that, although craftsmen worked less flexibly in the most highly organised workplaces, there had been fewer attempts by managers to promote flexibility at such workplaces in recent years. Although the association between trade union density and relatively less flexibility of working appeared strong and consistent at higher levels of density, we have again to emphasise that the association should be interpreted with caution. Because of the numbers involved it was not possible for us adequately to explore how far the association might have been an incidental consequence of some third influence. For instance, we had limited scope to explore the independent influence of size upon the association, although the analysis that was possible did suggest that the associations established between manual trade union density and our different measures of rigidity persisted independently of size.

Flexibility of working and the introduction of advanced technology

As we mentioned at the start of this section, we found strong and important associations between the introduction of advanced technology and flexibility of working. The measure of the use of advanced technology that best revealed these associations was the threefold division between different usages of microelectronics: first, those that had both process and product applications; secondly, those with process applications only; and thirdly, those with no application.

There were also important differences in relation to whether workplaces introduced advanced technical change, conventional technical change or organisational change in recent years, as we mentioned in Chapter IV. These we discuss a little later. First, Tables VII.10 – VII.13 summarise the results of our questioning on flexibility in relation to whether or not workplaces used microelectronics. Table VII.10 shows that there was little difference between places that had both process and product applications and others in the extent to which works managers felt there were limits on their freedom to deploy labour as they wished. This is important because all our evidence suggested that those workplaces tended to have been subject to the greatest and the widest range of

Table VII.10 Extent and nature of constraints on management prerogative to organise work, in relation to the use of advanced technology

Column percentages[a]

	Total	Process and product applications	Process application only	No application
Management constrained in organisation of work	34	31	39	30
Sources of constraint[a]				
Union agreements	20	22	22	17
Steward resistance	19	23	19	19
Lack of skills	18	15	21	17
Union member resistance	15	22	18	10
The technology used	14	17	13	14
Non-union member resistance	2	7	1	–
Base: all works managers				
Unweighted	*276*	*78*	*138*	*53*
Weighted	*76*	*14*	*35*	*25*

[a] See note F.

Table VII.11 Extent and nature of flexibility among production workers in relation to the use of advanced technology

Column percentages

	Total	Process and product applications	Process application only	No application
Any flexibility among production workers of the types listed[a]	99	99	100	98
Routine maintenance	*28*	*43*	*27*	*20*
Setting up machines	*66*	*88*	*70*	*51*
Cleaning	*89*	*90*	*87*	*96*
Transporting materials	*75*	*70*	*71*	*84*
Base: all works managers				
Unweighted	*276*	*78*	*138*	*53*
Weighted	*76*	*14*	*35*	*25*

[a] See note F.

Table VII.12 Extent and nature of flexibility among electricians in relation to the use of advanced technology

Column percentages

	Total	Process and product applications	Process application only	No application
Any flexibility among electricians of the types listed[a]	76	95	82	58
Work without semi-skilled mates	*55*	*59*	*61*	*48*
Some mechanical fitting	*55*	*71*	*57*	*42*
Work on electronic equipment	*67*	*67*	*80*	*49*
Base: all works managers				
Unweighted	*276*	*78*	*138*	*53*
Weighted	*76*	*14*	*35*	*25*

[a] See note F

changes. Tables VII.11 and VII.12 show that places operating advanced technology tended, in practice, to enjoy substantially higher levels of flexibility. So far as process workers were concerned, they were very much more likely to be normally involved in maintenance and the setting up of machines. In contrast, process workers in places with traditional technology tended to be involved more in cleaning and transporting materials. Electricians in places with advanced technology were substantially more likely normally to engage in tasks outside their traditional function. This was especially true so far as mechanical fitting work was concerned. In relation to this practice, as with so many of our other items measuring flexibility, the distinctions we made between the three types of workplace in relation to their use of advanced technology served as a clear continuum.

Table VII.13 shows a similar continuum in relation to the extent to which workplaces had sought a relaxation of traditional demarcations in recent years. Plants with both process and product applications of advanced technology were particularly likely to have made efforts to create multi-skilled craftsmen and enhanced craftsmen. Further analysis suggested that manufacturing plants with the most sophisticated advanced technology had taken this tendency furthest.

Table VII.13 Extent and nature of initiatives taken to promote flexibility in the previous four years in relation to the use of advanced technology

Column percentages

	Total	Process and product applications	Process application only	No application
One or more of the listed steps taken to promote flexibility[a]	43	62	41	35
Relaxation of production maintenance demarcations	*20*	*37*	*20*	*13*
Relaxation of craft/craft demarcations	*29*	*21*	*34*	*28*
Creation of new category of multi-skilled craftsman	*15*	*22*	*12*	*14*
Creation of new category of enhanced craftsman	*13*	*34*	*10*	*6*
Base: All works managers				
Unweighted	*276*	*78*	*138*	*53*
Weighted	*76*	*14*	*35*	*25*

[a] See note F.

Perhaps the most striking and instructive of our results regarding the association between the introduction of advanced technology and the breakdown of traditional demarcations arose from analysis of steps taken to promote flexibility in relation to the three distinct forms of major change that we identified at workplaces. In Chapter IV we described the distinctions we made between, first, places that introduced change involving advanced technology, in the previous three years; secondly, places that introduced change involving conventional technology; thirdly, places that experienced change independently of any new plant, machines or equipment; and fourthly, places that experienced no change of these three types. We showed how there were generally major differences between each of the three forms of change, especially in its acceptibility to the workforce. Our interviews with works managers further revealed that each of the forms of change was also associated in a distinctive way with measures to promote flexibility. Moreover, the accounts of works managers suggested even more strongly that the different forms of change provoked a very different type of response from workers and their representatives.

Table VII.14 Extent and nature of initiatives to promote flexibility in relation to types of major change in previous three years

	Advanced technical change	Conventional technical change	Organisational change	No change
One or more of the listed steps taken to promote flexibility[a]	52	(25)[b]	(83)	(23)
Relaxation of production maintenance demarcations	*27*	*(6)*	*(37)*	*(9)*
Relaxation of craft/craft demarcations	*32*	*(19)*	*(60)*	*(20)*
Creation of new category of multi-skilled craftsman	*16*	*(7)*	*(39)*	*(8)*
Creation of new category of enhanced craftsman	*19*	*(5)*	*(15)*	*(6)*
Base: all works managers				
Unweighted	*176*	*40*	*25*	*35*
Weighted	*37*	*12*	*7*	*20*

[a] See note F.
[b] See note B.

Table VII.14 shows how far workplaces sought relaxation in traditional demarcations in relation to the types of major change. It is apparent, first, as we reported earlier, that rumours of the death of productivity bargaining of the classical type at the end of the 1960s were greatly exaggerated. The clear implication in Table VII.14 is that many of the cases of organisational change independent of any technological innovation were in fact classical productivity agreements. Workplaces experiencing that type of change had, in the majority of cases, been seeking a relaxation in traditional demarcations as between craftsmen and frequently trying to establish a new category of multi-skilled craftsmen, and seeking a relaxation in traditional demarcations as between process workers and craftsmen.

Secondly, however, those places that introduced advanced technology were more than twice as likely as those involved in conventional technological innovation to have taken steps to promote flexibility. The contrast was particularly marked in relation to the promotion of new categories of craftsman, both the enhanced craftsman who combined many of the characteristics of the traditional manual craftsman but had, in addition, certain technical skills and also the multi-skilled craftsman with skills that were previously confined to two or more traditional craftsmen. Workplaces that had introduced advanced technology were also very likely to have been engaged in the relaxation of demarcations as between process and maintenance workers. In contrast, places that experienced change based upon the introduction of new conventional technology were no more likely than those which reported no major change to have been involved in the breakdown of traditional demarcations. It is apparent that the introduction of advanced technology was distinctively associated with the breakdown of the types of demarcation that have traditionally been the subject of productivity bargaining in large, unionised, manufacturing workshops.

When works managers' accounts of reactions to the three different types of change among the workers affected and their stewards were added to this picture, then it became very striking indeed. We reported in Chapter IV how change that involved the introduction of new plant, machinery or equipment was generally more popular than organisational change involving no new hardware. This was clear from the reports of both managers and union representatives. So far as works managers were concerned, we had their reports of the reactions of workers and stewards to the three different types of change, both when they first learned about it and, subsequently, at the time of interview. Works managers' accounts of initial reactions are shown in Table VII.15. It would be difficult to imagine a more dramatic difference between reported responses to news of the introduction of new technology compared with reactions to the prospect of organisational change. According to works managers, three

Table VII.15 Works managers' accounts of the initial reactions of manual workers and shop stewards to different types of proposed change

Column percentages[a]

	Reactions of manual workers affected			Reactions of shop stewards		
	Advanced technical change	Conventional technical change	Organisational change	Advanced technical change	Conventional technical change	Organisational change
Strongly in favour	41	(58)[b]	(–)	45	39	(–)
Slightly in favour	35	(16)	(24)	32	(21)	(16)
Slightly resistant	16	(21)	z(32)	18	(36)	(39)
Strongly resistant	7	(4)	(44)	2	(3)	(44)
Support score	+90	+100	–100	+100	+60	–110
Base: works managers who reported change						
Unweighted	*176*	*40*	*25*	*175*	*40*	*25*
Weighted	*37*	*12*	*7*	*37*	*12*	*7*

[a] See note C.
[b] See note B.

Table VII.16 Summary of works managers' accounts of the reactions of manual workers and shop stewards to different types of proposed change

Mean scores[a]

	Manual workers affected			Shop stewards		
	Advanced technical change	Conventional technical change	Organisational technical change	Advanced technical change	Conventional technical change	Organisational change
Initial reactions	+90	(+100)[b]	(−100)	+100	(+60)	(−110)
Subsequent reactions	+140	(+170)	(+70)	+150	(+150)	(+10)
Base: works managers who reported change						
Unweighted	*176*	*40*	*25*	*176*	*40*	*25*
Weighted	*37*	*12*	*7*	*37*	*12*	*7*

[a] See note I.
[b] See note B.

quarters of the workers involved welcomed the prospect of advanced technology. In contrast, three quarters of the relevant workers were hostile to the prospect of organisational change unaccompanied by any new machine. Forty-one per cent of workers were reported to be strongly in favour of particular advanced technological changes. Forty-four per cent of workers were at least equally strongly opposed to the idea of particular forms of organisational change. The general pattern reported for stewards was broadly similar to that for their members. Relating these results back to Table VII.14, it appears highly plausible that combining efforts to bring about more efficient work organisation with the introduction of new machinery was likely to make new working practices very much more palatable.

As well as having works managers' accounts of initial reactions we also had their judgements regarding subsequent feelings. The contrast between the two stages is summarised in Table VII.16. It shows the familiar shift towards a substantially stronger level of support that was apparent in the reports of both manual and non-manual stewards and which we discuss in the next chapter. The movement in reported response, however, was most marked in relation to organisational change. This revealed a similar degree of modification to the classical chronicling of workers' responses to proposals for productivity agreements that we also discuss in the next chapter(15). Initially, workers were said to be hostile to the idea of change in three quarters of the cases. Subsequently, when in most cases change had been implemented, there were reports of support for the change in three quarters of cases, though generally that support was described as modest. Moreover, it remained true that organisational change emerged as very much less popular than technological change, even after implementation. This contrast was particularly marked so far as the views of stewards were concerned. Their feelings remained evenly balanced between cases where they were reported to be supportive and those where they were felt to be hostile, even after implementation. We go on to consider the wider picture regarding levels of worker and trade union support for change more fully in the next chapter.

Footnotes

(1) A. Smith (1776), *The Wealth of Nations*, Penguin, Harmondsworth, 1974.

(2) F.W. Taylor (1911), *Principles of Scientific Management*, reprinted in *Scientific Management*, Harper Row, 1947.

(3) Charles R. Walker and Robert H. Guest, *The Man on the Assembly Line*, Harvard University Press, Cambridge, Mass., 1952; John H. Goldthorpe *et al.*, *The Affluent Worker: Industrial Attitudes and Behaviour*, Cambridge University Press, London, 1968.

(4) Robert Blauner, *Alienation and Freedom*, University of Chicago Press, Chicago, 1964.

(5) W.W. Daniel, 'Automation and the Quality of Work', *New Society*, Vol. 13, No.348, 1969.

(6) H. Braverman, *Labour and Monopoly Capital: the Degradation of Work in the Twentieth Century*, Monthly Review Press, New York, 1974.
(7) See, for example, J. Downing, 'Wordprocessors and the Oppresssion of Women', in Tom Forester (ed.), *The Microelectronic Revolution*, Basil Blackwell, Oxford, 1980.
(8) Peter Senker (ed.), *Learning to use Microelectronics: A Review of Empirical Research on the Implications of Microelectronics for Work Organisation, Skills and Industrial Relations*, Science Policy Research Unit, University of Sussex, 1984.
(9) F. Hersberg, *Work and the Nature of Man*, Staples Press, London, 1968.
(10) W. Allen, 'Half Time Britain on Half Pay', *Sunday Times*, 1 March, 1964.
(11) The first systematic analysis and account of such an agreement was provided by Allan Flanders, *The Fawley Productivity Agreements*, Faber and Faber, London, 1964.
(12) W.W. Daniel and Neil McIntosh, *Incomes Policy and Collective Bargaining at the Workplace*, PEP, No.541, 1973.
(13) Trades Union Congress, *New Technology and Collective Bargaining*, TUC, London, 1981.
(14) Government Social Survey, *Workplace Industrial Relations*, HMSO, 1968.
(15) W.W. Daniel and Neil McIntosh, *The Right to Manage*, Macdonald, London, 1972.

VIII Trade Union and Worker Support for Advanced Technical Change

Post-war British industrial strategy was not 'equal to the task of overcoming the massive inertial resistance to change manifested by the British industrial system — not least because it still failed altogether to address the problem of the trade unions, possibly the strongest single factor militating against technical innovation and high productivity.' Thus Correlli Barnett sums up his latest versions of a perennial diagnosis of Britain's industrial weakness. (1). Such diagnoses are commonplace in analyses of the performance of the British economy over the past 30 years. They identify trade union and worker resistance to change as a major reason for our comparatively poor economic performance. Special weight is attached to resistance from manual workers and the manual trade unions. Resistance to the introduction of new machines is singled out as a prime example of the way the attitudes of manual workers and their unions represent a generalised obstacle to innovation and efficiency. By implication, much less blame is attached to managers and employers. They are assumed to favour change. If the problem is worker resistance to change, then, by implication, managements are the allies of change and seek to bring it about.

In fact, although the idea that workers through their union structure resist technical change has become firmly rooted in conventional British wisdom, there is no well documented evidence to support that view(2). Previous PSI studies of the introduction of microelectronic applications into manufacturing industry have provided little hint that worker or trade union resistance represented an obstacle to the diffusion of the new technology(3). Indeed, Northcott found that managers in Germany and France were more likely to report worker resistance to advanced technology than counterparts in Britain(4).

Our present findings stand in even more marked contrast to the conventional wisdom than earlier evidence. We found that the reactions of workers and unions to technical change were generally very favourable. First, our interviews with general managers, personnel managers, works managers and shop stewards combined to show that the general response of the workers affected to specific advanced technological changes was overwhelmingly supportive. This was true for both manual workers and office workers affected by technical change. Secondly, where union officers were involved, their reactions tended to be even more supportive than those of the workers affected. Indeed, workers and union officers emerged from our analysis as so strongly in favour of technical change, in terms of their practical response to particular changes, that it would seem no longer possible to use the standard phrases *worker resistance to technical change* or *trade union resistance to technical change*. Instead, it would be more in accord with reality if commentators started to talk about variations in the strength of *trade union support for technical change* and *worker support for technical change*.

We looked briefly at worker reactions to different forms of change in Chapter IV. We demonstrated there, first, that there was generally a major difference in reactions to technical change compared with responses to organisational changes introduced independently of any new machinery or hardware. Both workers and trade union officers were substantially more strongly in favour of technical change than of organisational change. Secondly, there were also slight differences in reactions to different forms of technical change. Both workers and union representatives tended to support the introduction of new conventional technology more strongly than the adoption of new advanced technology. There was, however, markedly less difference between reactions to the two different forms of technical change than there was between both forms of technical change and organisational change. Again, this contrast was common to both manual workers and office workers. In this chapter, we focus upon reactions to the introduction of advanced technical change. In particular, we analyse managers' and shop stewards' accounts of how different categories reacted to the most recent change affecting them and we explore sources of variation in these reactions. As a preface to that analysis, we need to explain a little more the nature of the questions we asked.

So far as our principal management respondents were concerned, we asked them about the reactions of four different categories:

- the workers directly affected by the change;
- the foremen and supervisors of the manual workers concerned;
- the shop stewards or manual worker representatives;
- paid union officials outside the establishment.

For each of these categories our question to managers about reactions focussed upon the time *when they were bringing in the change*. This focus is important because previous research on the introduction of change showed that there tended to be different reactions at different stages, from the initial proposal for the change through the implementation stages to the operating stages(5). We were not able to take on board any of these complexities in our main management schedule, owing to pressure on space. In our interviews with shop stewards and works managers, however, we were able to distinguish between reactions at two different stages: first, when workers and stewards heard about the proposed change initially and, secondly, how they felt about it at the time of interview. Answers to these questions from manual shop stewards, non-manual stewards and works managers confirmed that there was a marked strengthening of the level of support between initial reactions and subsequent feelings. Accordingly, although we focussed the attention of principal management respondents upon the time when the change was being introduced, in asking them about the reactions of the different parties, it may well be that their answers were coloured by assessments of feelings at the time of interview. The reports of management respondents had to be interpreted in the light of accounts of shop stewards and of works managers and by the variations at different stages in the process that they revealed. These reports, as we show later in the chapter, suggested that the reactions of both stewards and the workers affected moved from modest support when the advanced technical change was first proposed to wholehearted support at the time of interview.

First, however, in this chapter, we analyse the accounts of principal management respondents of the reactions of different groups to advanced technical change affecting manual workers, when the change was being introduced. Secondly, we fill out the picture by adding the reports of manual shop stewards and drawing the distinction between initial reactions and subsequent feelings. Thirdly, we analyse comparatively the ways in which both managers and shop stewards described the reactions of different groups to the introduction of advanced technical change affecting office workers.

Managers' accounts of the reactions of manual workers and union officers to the introduction of advanced technical change

Managers reported that the workers directly affected were in favour of the change in three quarters of all the cases of advanced technical change affecting manual workers, identified in our interviews with primarily management respondents. In one half of these cases the workers were described as being strongly in favour. Where any resistance was reported, it was generally felt by managers to have been slight. The incidence of strong resistance was minute. One of the main influences upon the strength of support, according to managers' accounts, was size, as measured by the number of manual workers employed. This tendency was most apparent, however, in the contrast between reactions in large work places compared with those in smaller ones. The difference is most clearly revealed by the support score at the foot of Table VIII.1, which summarises and standardises managers' accounts of reactions. From the

Table VIII.1 Managers' accounts of the reactions of manual workers to the introduction of advanced technological change

Column percentages[a]

	All establishments	Number of manual workers at establishment				
		Fewer than 100	100-199	200-499	500-999	1000 or more
Manual workers' reactions						
Strongly in favour	37	40	29	40	24	20
Slightly in favour	37	37	44	31	39	31
Slightly resistant	19	18	16	20	30	42
Strongly resistant	2	2	3	4	5	2
Not stated	4	2	8	5	2	6
Support score[b]	+90	+100	+90	+90	+50	+30
Base: establishments with 25 or more manual workers and experiencing advanced technical change						
Unweighted	*458*	*106*	*70*	*121*	*104*	*57*
Weighted	*212*	*142*	*31*	*27*	*8*	*4*

[a] See note C.
[b] See note I.

full set of answers, it can be seen that the tendency for the strength of support to decline the larger the size of the establishment was largely brought about by a tendency for the level of slight resistance to increase the larger the workplace. Even in the largest workplaces there were very few cases where managers reported strong resistance from the workers affected by advanced technical change.

In Table VIII.2 we add to the picture managers' accounts of the reactions of shop stewards, full-time union officers and first-line managers. A crucial feature of the table is the contrast between the two halves. The first half is based upon all cases where advanced technical change was introduced affecting manual workers and 25 or more such workers were employed. Apart from the extent of the support from all groups, the most notable feature is the proportion of cases where managers reported that the reactions of union officers were not relevant because they were not involved. The proportion was over one third in the case of shop stewards and nearly three quarters so far as full-time officers were concerned. It was not simply that unions were not recognised in those cases. The proportions in the two categories fell only to 15 per cent and 64 per cent respectively when analysis was confined to workplaces where unions

Table VIII.2 Managers' accounts of the reactions of different categories associated with the introduction of advanced technical change affecting manual workers

Column percentages[a]

	Establishments having changes affecting manual workers				Cases in which the respective categories were involved			
	First-line managers	Manual workers affected	Shop stewards	Full-time officers	First-line managers	Manual workers affected	Shop stewards	Full-time officers
Strongly in favour	50	37	24	10	54	39	37	36
Slightly in favour	27	37	22	8	28	39	35	28
Slightly resistant	16	19	16	10	17	20	25	34
Strongly resistant	2	2	2	*	2	3	2	2
Not stated/not involved	5	4	36	72				
Support score[b]					+110	+90	+80	+60
Base: as column heads								
Unweighted	*458*	*458*	*458*	*458*	*438*	*432*	*367*	*141*
Weighted	*212*	*212*	*212*	*212*	*200*	*196*	*135*	*58*

[a] See note C.
[b] See note I for details of the calculation of scores. Throughout our analysis, support scores were derived from cases where respondents expressed a view (i.e. cases where the question was not applicable or the respondent was not able to answer were omitted). Accordingly the support scores are given under that part of the table where analysis is confined to those respondents for whom the question was relevant.

were recognised. The pattern is consistent with our earlier results showing that union officers were frequently not consulted in many instances where advanced technical change was introduced, even in places where unions were recognised.

The exclusion of union representatives from so many cases complicated the analysis of their reactions compared with those of their constituents. When we analysed the reactions of the respective parties basing percentages upon those cases where the responses of each were judged by managers to be relevant, we found the picture shown in the second half of Table VIII.2. The variation in the extent of support appeared to follow the path indicated by conventional views. Support was strongest among first-line managers, but according to managers' accounts, the level of support from foremen and supervisors was only slightly greater than that among the manual workers affected. The difference was largely accounted for by the fact that, in cases where managers felt that foremen supported the change, they were likely to describe that support as *strong* rather than *slight*. Managers reported some resistance from foremen in almost as many cases as they reported resistance from the workers affected.

When comparisons were made between all cases where the views of workers were reported and cases where the reactions of union officers were felt to be relevant, it appeared that there was less support for advanced technical change from shop stewards and even less from full-time officers, compared with the level of support from the workers affected. This impression, however, was seriously misleading. As mentioned earlier, the reactions of stewards were judged by managers to be relevant in only 62 per cent of cases and the reactions of full-time officers were felt to be relevant in only one quarter of cases (see Table VIII.2). Our analysis suggested that union officers were more likely to become involved where there were doubts or reservations among manual workers; this was particularly the case so far as full-time officers were concerned. Occasionally, full-time officers were called in by managers, as reported in Chapter VI, because they had some difficulty in introducing the change. Full-time officers were also more likely to be consulted by stewards where they had difficulties over the prospective change. In consequence, when we confined our analysis, first, to cases in which full-time officers were involved and, secondly, to cases in which stewards were involved, we found in Table VIII.3 a very different picture from that revealed by the second half of Table VIII.2. It was apparent that, even according to the accounts of managers, there was more support for change from full-time officers than from stewards or ordinary workers. The perspective of managers in these cases led them to conclude that levels of support from stewards were very similar to those from the workers affected.

A second way of looking at the contrast between the reported reactions of the different groups is by focusing upon the proportions who were described as resistant to change in the first half of Table VIII.2. Even though full-time officers were involved in the more difficult cases, resistance to change from full-time officers was reported in a lower proportion of all cases of change than from any other category, including first-line managers. Indeed, while some resistance was reported from full-time union officers in 10 per cent of all cases, it was reported from first-line managers in 18 per cent of all cases. Of course, this contrast largely arose from the fact that full-time officers were involved in so few cases, but the contrast does put into context the extent of resistance to change from the formal union structure.

A further major qualification that needs to be made to the comparisons in both Table VIII.2 and VIII.3 is that they were derived from the reports of managers. We show later in the chapter, when we compare the accounts of stewards with those of managers, that stewards themselves reported that they were substantially more strongly in favour of the change than the manual workers they represented. We also discuss later in the chapter the way in which negotiating stances may lead managers to judge that support for change among trade union representatives is less strong than it is. Notwithstanding any such tendency, however, the main result to emerge from this section is that managers attributed very substantial

Table VIII.3 Managers' accounts of the reactions of workers and union officers to advanced technical change in cases where union officers were involved

	Manual workers affected	Shop stewards	Full-time officers
Cases in which stewards were involved			
Average support score	+80	+80	+90
Per cent resistant	27	23	10
Unweighted base	*277*	*277*	*277*
Weighted base	*98*	*98*	*98*
Cases where full-time officers were involved			
Average support score	+50	+50	+60
Per cent resistant	37	37	37
Unweighted base	*141*	*141*	*141*
Weighted base	*58*	*58*	*58*

support to union officers for advanced technical changes that they recently introduced. In the cases where union officers were involved, managers reported that they received as much or more support from them than from the manual workers affected. This was so, despite any tendency which we discuss later in the chapter for trade union representatives to appear less strongly in favour of change than they might personally feel, owing to their negotiating positions. Before going on to review that tendency, when we compare stewards' accounts of their feelings with managers' reports of their reactions, we look at some of the main sources of variation in levels of support for change among manual workers and their representatives. The analysis was principally derived from managers' accounts but, unless otherwise stated, it was supported by the pattern of variation also revealed by the reports of manual shop stewards.

Variations in levels of support for change among manual workers and their representatives

When we examined patterns of variation in managers' reports of the reactions of the different manual groups to the introduction of advanced technical change, we found that they generally moved in concert. That is to say, in cases where there was a relatively low level of support among the manual workers affected, managers reported a similarly low level, in comparative terms, among shop stewards and full-time officers and, indeed, foremen and supervisors. This pattern had two implications. First, it provided evidence of overall congruence between the reactions of workers and of their representatives. Certainly it provided no hint, even in management eyes, that there was substantial dissonance between workers and their representatives regarding the introduction of advanced technical change. Secondly, in exploring patterns of variation, we were only able to focus upon the reactions of workers alone and it may be assumed that the reported reactions of other categories moved in similar ways, unless otherwise stated. We report analysis of variations in levels of support for change in relation to the size and sector of workplaces, the level of union organisation, the content of the change, and the manner in which it was introduced. The results of the analysis are summarised in Tables VIII.4 and VIII.5.

First, as we noted earlier, there was a marked fall in the level of support for change involving advanced technology in larger establishments, where 500 or more manual workers were employed. It is notable, however, that below that size there was little association between the level of support and the number of manual workers employed. It appeared to be the case that it was only when workplaces became very big that there was a substantial fall in the level of support for technical change. In relation to sector, the nationalised industries stood out as the category where

Table VIII.4 Variations in managers' accounts of worker reactions to change in relation to size, sector and union recognition

Mean scores and percentages

	Number of manual workers employed at establishment						
	All establishments	25-49	50-99	100-199	200-499	500-999	1000 or more
Support score (mean)[a]	+90	+90	+100	+90	+90	+50	+30
Proportion of cases where managers reported any resistance (per cent)	*21*	*23*	*14*	*19*	*23*	*35*	*44*

	Sector			
	Private manufacturing	Private services	Nationalised industries	Public services
Support score	+100	+100	+10	+120
Proportion of cases where managers reported any resistance (per cent)	*18*	*17*	*54*	*13*

	Trade union recognition			
	All establishments		Establishments with 50-199 manual workers	
	Manual union recognised	Union not recognised	Manual union recognised	Union not recognised
Support score	+90	+100	+100	+80
Proportion of cases where managers reported any resistance (per cent)	*24*	*11*	*16*	*16*

[a] See note I.

Table VIII.5 Variations in management accounts of worker reactions to change in relation to the form and content of the change

Mean scores and percentages

	Earnings			Manning			Job content		
	Increased	No change	Decreased	Increased	No change	Decreased	More	No change	Less
Support score (mean)[a]	+110	+90	−50	+120	+100	+60	+100	+90	+50
Proportion of cases where managers reported any resistance (per cent)	*15*	*21*	*72*	*11*	*19*	*33*	*19*	*21*	*29*

	Extent of worker involvement			Decision to introduce change taken at			
	Change negotiated	Consultation over change	No discussion	Independent establishment	Establishment	Joint decision	Higher level
Support score (mean)	+10	+90	+110	+120	+90	+120	+80
Proportion of cases where managers reported any resistance (per cent)	*52*	*24*	*18*	*11*	*19*	*17*	*27*

[a] See note I.

changes had enjoyed the least support from the manual workers affected. Managers reported some resistance from workers in the majority of cases. Generally, however, any resistance was slight and, in terms of our support score, was more than balanced by instances where support was strong. In consequence, the reaction overall remained slightly positive. Outside the nationalised industries a little more support for change was reported in the public sector than in private manufacturing and private services. The difference between public services and both parts of the private sector, however, was largely a consequence of differences in the characteristic size of workplaces, and, of course, the use of advanced technology affecting manual workers was rare in public services.

Secondly, and importantly, there was little sign of any association between the level of trade union organisation at workplaces and the extent of support for change. Certainly, as Table VIII.4 shows, when places of similar size were compared, there was no indication of support being any less strong in places where unions were recognised than in those where they were not. So far as union density was concerned, when places of similar size were compared, there was, if anything, a slight tendency for support to be stronger where trade union membership was high. Thirdly, the association between managers' accounts of workers' reactions and their separate accounts of the implications of change for earnings, manning and job interest were just as would be expected. Support was described as very strong when earnings or manning or job interest were increased as a result of the change. Support declined when the manning or job interest was reduced. There was resistance in those few cases where earnings fell as a result of the change.

Fourthly, in relation to the management level at which the decision to introduce change was taken, support was reported as strongest where the workplace was an independent establishment and the decision was necessarily taken by local management. Support was weakest where the workplace was part of a group and the decision was taken at a higher level. Where places were part of a group but the decision was taken locally, there was an intermediate level of support. The difference shown in Table VIII.5 became more marked when places of similar size were compared. An apparently anomalous category in that analysis is made up of those workplaces which were parts of groups, but when managers were asked whether the decision to introduce was taken by central or local management, they declined to opt for either and insisted that the decision was taken jointly by local and central managers. Generally, in survey analysis, people who opt for an intermediate position between two specified alternatives are not very interesting. In this instance, cases where managers insisted that decisions to introduce change were taken jointly, appeared to have a number of distinctive characteristics. One of them was certainly that they enjoyed a high level of support for change. Finally, in this section on variations in levels of support for change according to managers' accounts, we found, as Table VIII.5 shows, that there was a paradoxically strong tendency for level of support to decline the greater the extent of worker involvement in the change. We discussed the implications of this paradox in Chapter VI, where we showed that it was part of a general tendency on the part of managers to consult with manual workers or their representatives only when they were required to do so.

Manual stewards' accounts of their reactions and of the reactions of the workers affected

The addition of shop stewards' reports of reactions to advanced technical changes enabled us to add two very valuable further pieces to our jig-saw picture of worker and trade union support for change. First, as we explained at the beginning of this chapter, we were able to distinguish in our interviews with worker representatives between initial reactions to news of the proposed change and subsequent feelings about it at the time of interview. Secondly, the combination of both management and shop steward perspectives upon the same events provided a much more convincing account of actual reactions than either report would have given on its own.

First, as we mentioned earlier, research in the introduction of organisational change during the productivity bargaining boom of the late 1960s showed that there were marked differences in the reactions of the workers affected at different stages. There were changes in both their general evaluation of the proposed agreement and in the criteria according to which way they evaluated its implications(6). During the negotiating stage the focus tended to be the wage-work bargain and the size of the increases in pay being offered in relation to the changes in working practice being sought. Feelings tended to be sceptical about the proposed agreement or hostile to it. Following acceptance and implementation of the agreement, the feelings of workers about it contrasted markedly with their earlier reactions. The focus switched to the impact of the changes in working practice upon job content. Feelings moved towards support for the agreement, and the reasons given for approving it chiefly concerned the way in which work was made more interesting and satisfying as a result of an increase in responsibility, range of tasks and variety of tasks following upon the changes in working practice.

Table VIII.6 summarises the accounts of shop stewards in our present study of their own reactions and of those of the workers affected at the two stages we identified. The table also compares those accounts with the reports of managers in similar cases. The analysis shows, first, that, consistent with our previous research, there was a marked change in the reactions of both stewards and workers between the two stages. Secondly, however, that change consisted of movement from a slight balance in

Table VIII.6 Comparison of the accounts of managers and manual shop stewards of the reaction of workers and stewards to the introduction of advanced technical change

Column percentages[a]

	Managers' accounts			Manual stewards' accounts			
				Initial reactions		Subsequent feelings	
	Workers' reactions	Stewards' reactions		Workers' reactions	Own reactions	Workers' feelings	Own feelings
Strongly in favour	40	38	Strongly in favour	22	34	32	46
Slightly in favour	28	32	Slightly in favour	18	19	31	29
Slightly resistant	26	23	Mixed feelings	22	21	29	15
Strongly resistant	2	2	A bit doubtful	16	17	4	7
Not relevant/not stated	4	5	Very doubtful	22	10	3	3
			Not stated	–	*	–	*
Support score	+80	+80		+10	+60	+120	+130
Base: see footnote[b]							
Unweighted	*277*	*277*		*288*	*288*	*288*	*288*
Weighted	*98*	*98*		*101*	*101*	*101*	*101*

[a] See note C.
[b] The base for the analysis of managers' accounts in this comparison is those cases where managers gave accounts of the reactions of stewards and where stewards were interviewed. The base for the analysis of stewards' accounts is all cases where manual stewards reported advanced technical change in the previous three years.

favour initially, so far as workers were concerned, and a substantial balance in favour, so far as stewards were concerned, to an overwhelming balance in favour among both groups at the time of interview. The change did not entail, as had reactions to the productivity agreements, a movement from initial opposition to subsequent support. In this sense, worker responses to advanced technical change did appear to be different from their reactions to proposals for productivity agreements along classical lines. This view was strongly supported by our analysis in Chapter VII of works managers' accounts of reactions to different forms of change in manufacturing industry. We showed there, too, that feelings towards organisational change including productivity deals tended to move from initial opposition to subsequent support, while feelings towards technical change moved from modest support initially to very strong support subsequently.

Thirdly, Table VIII.6 shows that reports of reactions from stewards were broadly in line with the accounts given by managers in similar cases. This was particularly striking in view of the differences in the nature of the questions asked. The assessments of managers focussed upon just one period approximately at the mid-point between the stewards' separate assessments of initial and subsequent reactions. Both accounts suggested substantial support from both stewards and workers. There are two further points on the contrast between the separate reports of managers and stewards that warrant comment. First, stewards clearly felt that they favoured the changes substantially more strongly than the workers affected. The difference was particularly marked so far as initial reactions were concerned. This difference was not reflected in the assessments of managers from their perspectives. We discuss the implications of this type of variation more fully in our analysis of the reports from non-manual stewards and managers of the reactions to technical change of office workers and their union representatives. There was a more marked contrast in managers' and non-manual stewards' views of comparative levels of support from office workers and stewards, and the issues are more fruitfully reviewed in the context of changes affecting office workers.

Secondly, stewards' accounts of the initial reactions of the workers affected to the news of proposed advanced technical change do require us to add a note of qualification to our general argument that support for advanced technical change was widespread and strong among both manual workers and their union representatives. The accounts of manual stewards suggested that strong reservations about the prospect of advanced technical change were initially common in a substantial minority of cases. It appeared also, however, that these reservations were quickly overcome by events in the very large majority of cases.

Table VIII.7 Summary of manual stewards' accounts of their own and their members' reactions to the introduction of advanced technology

Mean scores

	Total	Private manu-facturing	Private services	Nation-alised industries	Public services
Initial reactions					
Manual workers affected (support score)[a]	+10	+20	(+40)[b]	(−70)	(+50)
Stewards' own reactions	+60	+80	(+60)	(+20)	(+60)
Subsequent feelings					
Manual workers affected	+120	+150	(+120)	(+40)	(+90)
Stewards' own reactions	+130	+170	(+120)	(+60)	(+110)
Base: establishments where manual stewards reported advanced technical change affecting manual workers					
Unweighted	*288*	*166*	*29*	*49*	*44*
Weighted	*101*	*49*	*16*	*23*	*14*

[a] See note I.
[b] See note B.

Variations in levels of support according to stewards' accounts

Tables VIII.7 and VIII.8 show variations in levels of support from manual stewards and the workers concerned in relation to both sector and size. The pattern was broadly in line with that revealed by the reports of managers. First, it is apparent that initial lack of enthusiasm was concentrated in the nationalised industries. Indeed, this was the one sector where the balance of feeling was hostile to the prospective change. Stewards reported that workers were not in favour of the proposed change in 53 per cent of the nationalised industry cases, and in one half of these cases workers were felt to have strong reservations. Ordinary workers were strongly in favour of the change from the start in only 8 per cent of the nationalised industry cases compared with nearly one quarter of cases overall.

Secondly, initial reservations among manual workers were also strongly concentrated in the largest establishments. Again it appeared that the critical cut-off point was the employment of 500 or more manual workers. Where that number were present on site, workers' initial reactions to proposed technical changes were unfavourable in more cases than they were favourable, according to stewards. In the largest workplaces there was initial support for the change, either slight or strong, in only 16 per

Table VIII.8 Summary of manual stewards' accounts of their own and their members' reactions to advanced technical change in relation to size

Mean scores[a]

	Total	Number of manual workers at establishment				
		Fewer than 100	100-199	200-499	500-999	1000 or more
Initial reactions						
Manual workers affected (support score)[a]	+10	+30	−20	*	−60	−90
Stewards' own reactions	+60	+80	+30	+50	+60	+30
Subsequent feelings						
Manual workers affected	+120	+140	+100	+100	+100	+100
Stewards' own reactions	+130	+130	+120	+140	+120	+130
Base: establishments where manual stewards reported advanced technical change affecting manual workers						
Unweighted	*288*	*51*	*49*	*74*	*71*	*43*
Weighted	*101*	*56*	*20*	*17*	*6*	*3*

[a] See note I.

cent of cases. This compared with 40 per cent in all cases of advanced technical change affecting manual workers where we had reports of reactions from stewards.

The support of office workers and non-manual unions for advanced technical change

The view that we quoted at the beginning of this chapter, which suggested that resistance to technical change among workers, expressed through trade unions, has been one of the main reasons for the comparatively poor performance of the British economy, focussed upon manual workers and their representative bodies. This has been the general form of the diagnosis. As we have indicated earlier in this book, technical change in the office has historically been less frequent and less substantial than on the shop floor. It was not until the advent of the mainframe computer early in the 1960s that technical change in the office began to receive attention as a matter of interest in its own right(7). At the same time, the growth of white-collar trade unionism in the private sector began to make managers fear that they would begin to experience organised resistance to change from a category of employees whose compliance they had always taken for granted. Our account of reactions to technical change on the shop floor in the first half of this chapter showed that any view which characterised manual workers as hostile and resistant to technical change and office workers as deferential and supportive, was clearly deficient in the first of the propositions. It appeared that so widespread was support for technical change among manual workers that it would have been difficult for it to have been exceeded by office workers. In the event, we found that the level of support from office workers and non-manual union representatives for the introduction of word processors and computers into offices was very similar to the support from their manual counterparts for advanced technical change on the shop floor. The distinctive features of non-manual reactions were twofold. First, the reactions of union officers to technical change in the office were most relevant in the public sector because the recognition of non-manual unions was very much more common in that sector. Secondly, there was a marked contrast in the public sector between managers' and non-manual stewards' accounts of levels of support for change among the office workers affected by the change compared with those of non-manual shop stewards. Accordingly, our analysis of levels of worker and trade union support for technical change in the office largely focuses upon those distinctive features relative to change on the shop floor. Initially, however, we look at managers' accounts of the reactions of office workers to the introduction of word processors and computers compared with their reports of responses to advanced technical change on the shop floor.

As Table VIII.9 shows, managers' separate accounts of reactions at the

Table VIII.9 Comparison of managers' accounts of the reactions of manual workers and office workers to the introduction of advanced technology

Column percentages[a]

	All relevant establishments		Establishments where advanced technical change affected both categories of employee	
	Manual workers	Office workers	Manual workers	Office workers
Strongly in favour	37	38	29	35
Slightly in favour	37	40	42	41
Slightly resistant	19	14	20	19
Strongly resistant	2	2	4	1
Not stated	4	6	5	4
Support score[b]	+90	+110	+80	+90
Base: as column heads[c]				
Unweighted	*458*	*977*	*326*	*326*
Weighted	*212*	*500*	*78*	*78*

[a] See note C.
[b] See note I.
[c] See note to Table V.5.

two different levels suggested that there was marginally even greater support for change among office workers. The difference was chiefly accounted for, however, by the tendency for managers to be a little more likely to report that office workers were slightly in favour and a little less likely to report that they were slightly resistant. This pattern emerged from a comparison of managers' reports in all cases of advanced technical change in offices with all cases of advanced technical change affecting manual workers. When we confined the analysis to those workplaces which introduced separate advanced technical changes affecting both categories, there was little difference in the comparison (see the second part of Table VIII.9).

Just as we did for changes affecting manual workers, we asked managers not only about the reactions to office changes of the office workers affected but also, where applicable, about the responses of their union representatives and supervisors. Table VIII.10 summarises management accounts of the reactions of the four categories. It may be compared with Table VIII.2 which showed the responses of their manual counterparts. Superficially, the comparison suggests that there was a more marked dif-

Table VIII.10 Managers' accounts of the reactions of different categories associated with the introduction of word processors or computers affecting office workers

Column percentages

	Establishments having changes affecting office workers				Cases in which the respective categories were involved			
	First line managers	Office workers affected	Shop stewards	Full-time officers	First line managers	Office workers affected	Shop stewards	Full-time officers
Strongly in favour	55	38	8	2	60	40	25	18
Slightly in favour	24	40	14	4	26	43	45	38
Slightly resistant	9	14	7	4	10	15	24	36
Strongly resistant	3	2	2	1	3	2	5	7
Not stated/not involved	9	6	69	88	–	–	–	–
Support score[a]					+130	+110	+60	+30
Base: as column heads								
Unweighted	977	977	977	977	920	938	534	192
Weighted	500	500	500	500	455	465	153	58

[a] Throughout our analysis support scores were derived from cases where respondents expressed a view (i.e. cases where the question was not applicable or the respondent was not able to answer were omitted). Accordingly the support scores are given under that part of the table where analysis is confined to those respondents for whom the question was relevant.

ference in the reactions of the different office groups to change compared with responses on the shop floor. It appears that, while the supervisors and office workers affected were more favourably disposed to change in the office than their manual counterparts, shop stewards and full-time union officers were less supportive. This superficial impression is misleading, however, for two reasons.

First, we saw in relation to manual workers that the impression, from managers' accounts, that stewards and full-time officers were less strongly in favour of change than ordinary workers was a consequence of the fact that union officers were not involved in many of the more straightforward cases of advanced technical change. In those instances where stewards and full-time officers were involved, stewards were felt by management to be as supportive of change as the manual workers involved and full-time officers were felt to be more supportive. The implication was that stewards and, to a greater extent, full-time officers became involved in cases only when the initial reactions of the workers affected were doubtful. They then tended to take views as supportive or more supportive of change, even according to the accounts of managers. Manual stewards themselves reported that they supported the changes substantially more strongly than the workers affected, especially at the initial stage. Table VIII.11 shows that the reactions attributed by managers to the different non-manual categories also tended to be much

Table VIII.11 Managers' accounts of the reactions of different categories to the introduction of office changes involving word processors or computers

Mean scores[a]

	All office cases	Cases where stewards were involved	Cases where full-time officers were involved
Reactions of office workers affected	+110	+80	+70
Reactions of stewards	+60	+60	+40
Reactions of full-time union officers	+30	+30	+30
Unweighted	*977*	*534*	*192*
Weighted	*500*	*153*	*58*

[a] See note I.

closer when analysis was confined to cases in which they were all involved. It remained true, however, unlike the pattern for manual workers, that officers were felt by managers to have been less strongly in favour of the change than the workers affected, even when the comparison was confined to instances where all three parties were involved.

The second and distinctive aspect of the reactions attributed to different categories of office staff was the contrast between the patterns that characterised different sectors of employment. Table VIII.12 summarises some of the principal features. It is immediately apparent that there were dramatic differences between the public and private sectors in the extent to which union officers,both shop stewards and full-time officials, were involved in the introduction of change in the office. At one extreme, in the nationalised industries, the involvement of shop stewards was almost universal and full-time officers were involved in nearly half the cases. At the other extreme, in private services, the reactions of shop stewards were judged to be relevant in about one case out of every seven, and full-time officers were involved in only one major office change out of every 17. Even within the private sector there were major differences between manufacturing industry and services. The involvement of stewards was nearly twice as common in the introduction of word processors and computers in the offices of private manufacturing industry. In the public sector, union involvement was very common in local and central government, but only half as common as in nationalised industries.

As would be expected, the relative frequencies with which union officers were involved in the different sectors had major implications for the extent to which any resistance from union officers was reported (see the foot of Table VIII.12). The pattern in the private sector was broadly similar to that identified for applications of advanced technology affecting manual workers. Largely because the involvement of full-time officers was so rare, resistance from them was substantially less common in the private sector, according to the reports of managers, than resistance from either the workers affected or from their first-line managers. Any resistance to office change from shop stewards was also very infrequent in the private sector, partly, again, because stewards were involved in so few instances. Here it should be emphasised that the figures at the foot of Table VIII.12 include all cases where any resistance, slight or great, was attributed by managers to the groups. In most cases where any resistance was reported, it was felt by managers to be slight.

In the public sector, partly because the involvement of union officers was so much greater, some resistance from them to the introduction of change in offices was more frequently reported by managers. Nationalised industries, especially, provided the exceptional case where managers reported more resistance from both stewards and full-time officers than from the office workers affected by the change. Nevertheless,

Table VIII.12 Proportion of office changes in which trade union officials were involved and where there was any resistance from different categories, according to managers

Percentages

	All establishments	Private manufacturing	Private services	Nationalised industries	Public services
(a) Proportion of cases where stewards were involved	31	27	15	94	51
(b) Proportion of cases where full-time officers were involved	12	7	6	46	21
(c) Proportions of all cases where any resistance (slight or great) was reported from the listed categories					
First-line managers	12	10	9	12	20
Office workers affected	16	12	14	20	22
Shop stewards	9	6	3	24	20
Full-time officials	5	3	1	22	12
Base: cases of change involving word processors or computers affecting office workers					
Unweighted	*977*	*348*	*255*	*105*	*269*
Weighted	*500*	*121*	*231*	*24*	*124*

even in that instance, the balance of view among union officers was favourable to change, as shown in Table VIII.13, where we summarise the reactions attributed to the different categories by managers. In each instance the base for the score is the cases in which each particular category was involved. It is apparent that, in the private sector, there was a similar tendency to that apparent among manual workers for the reactions of union officers to be similar to those of their members when the comparison was limited to the cases in which they were engaged. In contrast, the pattern for the public sector showed a marked tendency for managers to attribute less support to stewards than to the office workers involved, and even less to full-time officers. This was so, even when the analysis was confined to examples of the introduction of advanced technological change in offices where full-time officers were involved. Public services provided the very exceptional case where, according to management accounts, the reactions of both stewards and full-time officials were both, on balance, very slightly favourable. This pattern, however, which emerged from the reports of managers, was very much placed in perspective when we added the accounts of non-manual shop stewards.

Table VIII.13 Accounts of managers in different sectors of the reactions to office change of different categories (a) in cases where stewards were involved (b) in cases where full-time officers were involved

Support scores[a]

	Total	Private manu-featuring	Private services	Nation-alised industries	Public services
(a) **Cases where stewards were involved**					
Reactions of office workers	+80	+100	+90	+90	+50
Reactions of stewards	+60	+90	+90	+70	+30
Reactions of full-time officers	+30	+50	+100	+20	−20
Unweighted base	*534*	*189*	*71*	*95*	*179*
Weighted base	*153*	*33*	*36*	*23*	*63*
(b) **Cases where full-time officers were involved**					
Reactions of office workers	+70	+90	+100	+100	+50
Reactions of stewards	+40	+50	+100	+50	*
Reactions of full-time officers	+30	+50	+100	+20	−20
Unweighted base	*192*	*43*	*32*	*51*	*66*
Weighted base	*58*	*8*	*14*	*11*	*26*

[a] See note I.

Non-manual shop stewards' accounts of reactions to the introduction of word processors and computers

Our picture of the reactions of shop stewards and the office workers affected to the introduction of change in offices was completed by the addition of the reports by stewards themselves of those reactions. These reports added a new and intriguing feature to the jig-saw. As was the case in our interviews with manual stewards, we were able in our discussions with non-manual representatives to distinguish, first, between their own reactions and those of their members when they initially heard about the proposed change and, secondly, how they felt about it later, at the time of interview. By that stage the change was implemented in most instances. As the second half of Table VIII.14 shows, stewards reported that they and the office workers affected were favourably disposed to the change from the start. But, according to the stewards' accounts, they supported the change substantially more strongly than the workers affected. For instance, 37 per cent of stewards said that they were strongly in favour from the start, while only one quarter said that the workers affected were

Table VIII.14 Comparison of managers' and non-manual representatives' accounts of reactions to the introduction of word processors or computers

Column percentages[a]

	Manager's account			Non-manual stewards' accounts			
				Initial reactions		Subsequent feelings	
	Workers' reactions	Stewards' reactions		Workers' feelings	Own feelings	Workers' feelings	Own feelings
Strongly in favour	29	15	Strongly in favour	25	37	34	50
Slightly in favour	42	35	Slightly in favour	19	24	30	22
Slightly resistant	18	17	Mixed feelings	20	14	26	20
Strongly resistant	3	3	A bit doubtful	23	14	2	5
Not relevant/not stated	8	29	Very doubtful	11	10	7	4
			Not stated	–	1	1	*
Support score	+80	+60		+30	+80	+101	+140
Base: see footnote[b]							
Unweighted	*485*	*485*		*520*	*520*	*520*	*520*
Weighted	*152*	*152*		*201*	*201*	*201*	*201*

[a] See note C.
[b] See note to Table VIII.6.

strongly in favour. Even in cases where there were doubts initially, however, it appeared that they were largely dispelled by experience of the change. At the time of interview, support for the change had increased among both office workers and stewards. The movement in opinion appeared to be especially marked among ordinary office workers.

As the questions asked of stewards about reactions to the change were different from those asked of managers, it was not possible to make strict comparisons between them. Nevertheless, it is clear from the first part of Table VIII.14 that, overall, the respective reports of managers and stewards were broadly consistent. According to their separate accounts, there was substantial support for the change among both office workers and their stewards. One noteworthy feature of managers' accounts was the extent to which they reported that the reaction of stewards to the introduction of change in the office was not relevant in a substantial minority of cases, even in workplaces where we interviewed non-manual stewards. A second feature of managers' perceptions, compared with those of stewards, was that they attributed more support to the workers affected than to the stewards, while the stewards said that they were more strongly in favour than their members. This difference took on added interest when we looked at stewards' accounts of their own reactions and those of office workers within different sectors, and compared the picture revealed with that suggested by the reports of managers which we discussed earlier. Table VIII.15 summarises the picture according to stewards. Table VIII.13 showed the equivalent picture derived from managers' reports.

It is clear from Table VIII.15 that the difference between the reactions of stewards and workers to the introduction of technical change was most marked in the public sector, according to the stewards' reports. This was also the case so far as the managers' accounts were concerned. The striking feature of the comparison between Tables VIII.15 and VIII.13, however, is that, while stewards in the public sector saw themselves as having been very much more strongly in favour of change than their members, managers perceived workers as having been substantially more strongly in favour of the change than stewards. One conclusion is quite clear from the comparison. There was very much greater dissonance between stewards' perceptions of the stances adopted by the parties to office changes and managers' perceptions of those stances in the public than in the private sector. And white-collar unionism was very much more involved in the introduction of change in the public sector. So any distortions arising from dissonance in perceptions will have had substantial implications.

There are, of course, very good reasons why managers and stewards might develop different impressions of the positions adopted by the workers affected by the change and the stewards representing them.

Table VIII.15 Non-manual trade union representatives' accounts of reactions to changes involving word processors or computers (a) when first suggested; (b) at the time of interview

Support scores[a]

	Total	Private manu-featuring	Private services	Nation-alised industries	Public services
Initial reactions					
Office workers affected	+30	+30	+50	+10	+30
Stewards' own reactions	+80	+60	+70	+100	+80
Subsequent feelings					
Office workers affected	+110	+120	+90	+120	+110
Stewards' own reactions	+140	+130	+100	+150	+140
Base: establishment where non-manual stewards reported change involving word processors or computers.					
Unweighted	*520*	*169*	*66*	*78*	*207*
Weighted	*201*	*33*	*36*	*25*	*107*

[a] See note I.

Some office workers might be inclined to defer to managers and conceal from them any qualms that they had about the prospective change, and to take their doubts and criticisms to their stewards. While supporting the change themselves, stewards might feel that they had to represent to managers the reservations of their members. Equally, stewards might sometimes judge that they were likely to win a better deal for their members out of any negotiations over the change if they put over to managers a picture of a workforce reluctant to change. On such grounds, the dissonance between the accounts of office workers' and stewards' reactions to change given by managers and stewards in the public sector was not at all surprising. The surprising feature is the extent of congruence between the views of managers and stewards about office change in the private sector and about change affecting manual workers in all sectors. Certainly, the picture we have identified regarding office change in the public sector suggested an unusually high lack of accord in that sector between both the views of stewards and their members and between stewards and managers. In other words, first, the gap between the reported views of stewards and ordinary workers was greater in the public than the private sector. This was true according to the accounts of both stewards and managers. Secondly, while stewards reported that they were substantially more in favour of the changes than their members, managers felt that they were substantially less strongly in favour. It is difficult to see how such high levels of dissonance could facilitate the smooth introduction of change.

Variations in support for technical change among office workers

Patterns of variation in levels of support for technical change among office workers and their representatives were dominated by the contrasts between the reports of managers and stewards in the different sectors. The fact that trade union involvement in office change was concentrated in the public sector and that there was so much dissonance between the separate accounts of managers and stewards in that sector made it a little difficult to explore other sources of variation by looking at the reports of just one party. On the other hand, although trade union involvement in office changes was more common in the public sector, the introduction of word processors and computers in offices was less common in that sector. Partly in consequence of this pattern, non-manual stewards were involved in only about one third of all advanced technical changes that we identified in offices, according to managers. It is worthwhile having a brief look at sources of variation in management reports of the extent of office workers' support for change, and then checking how far that picture was modified by the minority of cases where we also had non-manual representatives' accounts.

As was the case with advanced technical change affecting manual workers, there was a tendency for managers to report less strong support for the change the larger the number of non-manual workers employed at the workplace. But despite this tendency, it remained the case that, even in places employing 1,000 or more non-manual workers, there was some resistance, according to managers, from the office workers affected in only 22 per cent of cases, and in the large majority of these cases the resistance was described as slight. The association between declining shop steward support for the change and size was even stronger than the association relating to the office workers affected. Both associations were consistent with the accounts of shop stewards. According to the reports of stewards, too, they supported particular changes less strongly themselves and there was less support from their members in larger workplaces. The trend was most pronounced, however, in relation to stewards' accounts of feelings at the time of interview rather than in relation to their initial reactions. A possible implication is that the distinctive reactions of workers and stewards in larger workplaces to the technical changes had more to do with the manner in which they were introduced in larger workplaces, or the form they took, rather than any characteristic predispositions among office workers employed in larger workplaces to respond unfavourably to technical change.

There were slight tendencies for managers to report less support for change from the office workers affected in places where unions were recognised than in those where they were not. This was certainly the case so far as places employing larger numbers of non-manual workers were concerned. The association, however, was reversed in smaller places, suggesting that it was by no means a simple one. Moreover, as we have already analysed in some detail, in cases where unions were recognised, the reports of non-manual stewards often revealed a different picture from that painted by managers. Some of the most intriguing patterns of variation revealed by managers' accounts of reactions, however, lay in the contrasts between the reactions of the workers affected and those of non-manual union representatives in relation to the implications of the change for staffing, earnings and intrinsic job interest. For instance, the response attributed to both stewards and full-time officers was very strongly associated with the impact that the change had upon staffing, but this was less true for the office workers affected. In contrast, the reactions of the workers themselves were strongly associated with the implications of the change for intrinsic job interest. The reactions attributed to stewards mirrored this pattern, but the reactions attributed to full-time union officers was not at all associated with the impact of the change upon job interest. Similarly, as would be expected, the support of the workers affected was enhanced in cases where earnings rose as a result of the change. But there was not the same positive association between increases in pay and the reactions attributed to stewards and full-time officers. The implication of these patterns is that non-manual union officers were primarily concerned about the impact of the change upon employment and less concerned about its implications for pay so long as it was not reduced, or for intrinsic job interest. The reactions of the office workers affected were, in contrast, strongly influenced by the implications of the change for job interest and support was heightened where earnings were increased, but they had less concern about the implications for employment.

Footnotes

(1) Correlli Barnett, *Audit of War*, Macmillan, London, 1986.

(2) Peter Senker (ed.), *Learning to use Microelectronics: A Review of Empirical Research on the Implications of Microelectronics for Work Organisation, Skills and Industrial Relations*, Science Policy Research Unit, Sussex University, 1984.

(3) All three of Northcott's PSI surveys of manufacturing industry in 1981, 1983 and 1985 found that worker and union resistance ranked very low in the difficulties experienced by managers in introducing new technology based on microelectronics.

(4) Jim Northcott *et al.*, *Microelectronics in Industry: An International Comparison: Britain, France and Germany*, PSI, No.800, 1985.

(5) W.W. Daniel and Neil McIntosh, *The Right to Manage*, Macdonald, London, 1972.

(6) *Ibid.*

(7) Enid Mumford and Olive Banks, *The Computer and the Clerk*, Routledge and Kegan Paul, 1967

IX Advanced Technology and Loss of Jobs

The idea that new machines or methods would lead to job loss and unemployment has haunted the people affected by technical change since the start of the industrial revolution. The industrial application of the microchip revived in a new and dramatic way what had been a rumbling debate over the years. Here was a major new source of power heralding the development of a second industrial revolution. It promised opportunities to replace manpower in an unprecedented way. Some predictions of the likely impact upon jobs in the early stages were very alarming(1).

These predictions, however, have not been fulfilled or certainly not yet. In practice, many of the productivity gains promised for the new technology did not materialise; or users took advantage of the increased productivity to do more work with the same resources; or greater efficiency enabled users to compete more effectively and grow(2). Studies of the extent of job loss associated with the introduction of advanced technology showed that, overall, it was very modest(3). We are able usefully to supplement available information from our present survey, but we are certainly not able to answer all the questions raised in relation to the implications of advanced technology for employment. Our present information enabled us to focus upon three sets of questions: first, how far the introduction of advanced technology into the office or on to the shop floor led to immediate reductions in staffing or manning in the sections directly affected; secondly, how far general movements in the size of the establishment's workforce over the previous four years were associated with the extent to which it used advanced technology; and thirdly, how far the patterns of recruitment and displacement at the workplace were associated with its use of new technology. We found that there was a very slight tendency for job loss to be associated with the use of advanced technology, but that its use ranked very low among the whole range of reasons for reductions in the size of workforces. The introduction of advanced technology appeared to have much more impact upon the categories of people employed than upon the numbers employed. The questions left unanswered by our information were twofold. First, we were not able to assess how far any reductions in manpower associated with the introduction of new technology in the private sector might have been even greater in its absence, owing to loss of competitiveness. Secondly, we could not tell how far increases in productivity which resulted from the introduction of new technology in particular workplaces were translated into job growth elsewhere, outside those workplaces. We were not able to explore the possibility of such wider, indirect effects with our present data(4). Our information was limited to the extent to which reductions in manning, movements in the size of the workforce and patterns of recruitment were associated with the introduction of new technology within establishments. Nevertheless, the impact that the introduction of new technology has upon jobs within particular workplaces remains of special interest. In particular, the effect of change upon jobs at their own workplaces is one of the chief concerns of employees and their representatives in reaching their view about the desirability of the change(5).

Impact of advanced technology upon the sections of manual workers affected

In the majority of cases the introduction of advanced technology on the shop floor did not have any short-term consequences for the number of manual workers employed in the section or sections affected. This was the case according to the reports of both management and manual shop stewards as shown in Table IX.1. (In fact, as Table IX.1 compares the accounts of managers and shop stewards, the figures are based upon

Table IX.1 Comparison of managers' and manual stewards' accounts of the impact of advanced technical change on manning in section(s) affected

Column percentages[a]

	Manual union recognised		Workplaces where both managers and stewards were interviewed about an advanced technical change	
	All managers	All shop stewards	Managers	Stewards
Manning levels in affected section(s)				
Increased	11	8	12	7
Decreased	22	24	26	25
No change	67	68	62	67
Base: see footnote [b]				
Unweighted	*405*	*288*	*205*	*205*
Weighted	*160*	*101*	*72*	*72*

[a] See note C.
[b] See note to Table VI.7.

Table IX.2 Impact of advanced technical change on manning in section or sections affected

Column percentages

	All establishments	Number of manual workers					
		25-49	50-99	100-199	200-499	500-999	1000 or more
Manning levels in affected section(s)							
Increased	11	6	18	14	10	15	9
Decreased	19	6	12	37	44	38	33
No change	70	87	70	48	46	47	54
Not stated	*	*	–	1	–	*	4
Base: establishments with 25 or more manual workers experiencing advanced technical change.							
Unweighted	*458*	*50*	*56*	*70*	*121*	*104*	*57*
Weighted	*212*	*91*	*51*	*31*	*27*	*8*	*4*

workplaces where manual unions were recognised and, as we show later, manpower reductions more frequently followed technical change in such workplaces, as may be seen by comparing Table IX.1 with the total column in Table IX.2.) In places where manual unions were recognised, however, both managers and stewards reported independently that there was no change in about two thirds of all cases. The similarity between the separate reports of stewards and managers was remarkable. In the minority of cases where change was reported, it was much more common for manning levels to be decreased. But increases were reported in about one third of the cases where there was change, according to managers, and in about one quarter of the cases, according to stewards.

When we extended the analysis from places where manual unions were recognised to all workplaces where managers reported advanced technical change, the proportion of workplaces where the introduction of advanced technology led to reductions in manning in the affected sections fell to 19 per cent (see Table IX.2). Table IX.2 also shows, however, that there was some tendency for reductions in manning to be more common the larger the size of the workplace. In consequence, although only 19 per cent of workplaces experiencing technical change were subject to consequential reductions in manning, 34 per cent of all manual workers covered by our survey were employed at these establishments.

In relation to workplaces of all sizes, changes in manning resulting from the introduction of new technology were most common in private manufacturing and least common in public services (see Table IX.3).

Table IX.3 Impact of technical change on manning in section(s) affected in relation to sector

Column percentages[a]

	All establishments	Manufacturing	Private services	Nationalised industries	Public services
Manning levels in affected sections(s)					
Increased	11	16	7	8	2
Decreased	19	23	11	14	17
No change	70	61	82	77	81
Not stated	*	*	–	*	*
Base: establishments with 25 or more manual workers and experiencing advanced technical change					
Unweighted	*458*	*286*	*59*	*63*	*50*
Weighted	*212*	*116*	*44*	*26*	*26*

[a] See note C.

Table IX.4 Impact of technical change on manning in section(s) affected at larger workplaces in relation to sector

Column percentages[a]

	Establishments with 200 or more manual workers		
	Private manufacturing	Private services	Public services
Manning level in affected section(s)			
Increased	15	–	(17)
Decreased	40	(49)	(28)
No change	45	(51)	(54)
Not stated	1	–	–
Base: establishments with 200 or more manual workers and experiencing advanced technical change			
Unweighted	*191*	*23*	*23*
Weighted	*25*	*4*	*3*

[a] See note C.

The balance of those changes that were made in public services, however, were least favourable to employment. When we compared the pattern for different sectors among workplaces employing larger numbers of manual workers, where reductions were generally more common, we found the pattern shown in Table IX.4. It is especially notable that the private service sector was most likely to make reductions as a result of advanced technical change and least likely to make increases. The pattern is consistent with our analysis later in the chapter, which shows that loss of employment was most strongly associated with advanced technology in private services.

Reductions in manning in relation to union organisation

There was no sign that the level of manual union organisation at workplaces led to more favourable short-term outcomes, so far as employment was concerned. In places where unions were not recognised, more employers increased than decreased manning as a result of the change. Where manual unions were recognised, the opposite was the case. This contrast remained when we took into account differences in the size of workplaces. So far as union density was concerned, there was some suggestion that change in the level of manning was more common the higher the proportion of manual union workers, but there was no suggestion that change was more favourable to employment. Indeed, places where there was 100 per cent union membership tended to experience the least favourable outcomes in terms of changes in manning.

The tendency for those places where there was full manual union membership to be more likely to experience reductions in manning raised the possibility that, where there were reductions in manning, workers might more frequently receive increases in pay by way of compensation or as an inducement to accept change. Overall, however, there was no sign of any such association, when we analysed the extent of reductions in manning in relation to changes in earnings and intrinsic job interest arising from the change. Generally, in places where the innovation led to increases in earnings it was also more likely also to result in increases in manning than in reductions. In cases where earnings remained the same or fell, manning was also reduced more frequently. The pattern relating to assessments of the implications of the change for intrinsic rewards, however, was different. In cases where managers reported that the changes made jobs more interesting, it was much more likely than average for manning to have been reduced. In cases where a fall in intrinsic interest was reported, the balance of change in manning was favourable.

Scale of reductions in manning resulting from the introduction of advanced technology in manufacturing industry

So far as the overall impact of advanced technology upon levels of manning in sections of manual workers was concerned, we had only managers' and stewards' accounts of whether it had resulted in an increase, a decrease or no change. We had no measure of the scale of any increase or decrease. At those workplaces in manufacturing industry where there were personnel managers, however, we also had interviews with works managers. We had time in our discussions with them to collect more detail about any major changes at their plants. The extra detail included the number of people employed in sections affected by change before its introduction and the number employed at the time of interview. This enabled us to calculate the proportionate scale of any decreases in manning resulting from the changes. We also asked how any reduction in manning was brought about. We have to emphasise that this analysis is confined to changes in manufacturing industry and to those workplaces in manufacturing that had both a personnel function and a works management function. They tended to be the larger manufacturing establishments. On the other hand, the introduction of advanced technology affecting manual workers was concentrated in manufacturing industry and in the larger manufacturing plants. Hence our information from works managers provided a useful, if partial, indication of the scale of any changes in manning resulting from particular changes involving advanced technology in manufacturing industry, and of how those changes were brought about.

According to the reports of works managers, the introduction of advanced technology led to reductions in manning in the sections affected in about one quarter of cases and to increases in 17 per cent of cases. In the majority of cases, there was no change. These proportions were very similar to the pattern suggested by principal management respondents in all manufacturing establishments that we covered (see p.XX for details). The incidence of increases in manning from technological change meant that we had only 19 cases (unweighted) where works managers reported an increase. This number was too small for any real analysis, but works managers' reports suggested that, on average, about 15 or so people were employed in the sections which experienced increases in manning as a result of advanced technical change, and that number was increased by about one third as a result of the change.

We had more cases (59 unweighted) where works managers reported a reduction in manning. The difference in the unweighted number of cases where there was a decrease and the number where there was an increase was even greater than might have been expected from the respective percentages, because decreases were more common in larger plants and increases more common in smaller ones. Table IX.5 shows that, in manufacturing workplaces subject to reductions in manning as a consequence of advanced technical change, there tended to be about 50 people employed in the section most affected by the most recent change and their number tended to be reduced by over 40 per cent as a result of the change. The analysis suggests that the introduction of advanced technology had more damaging implications for the employment of manual workers in manufacturing industry than was implied by our earlier re-

Table IX.5 Works managers' accounts of the scale of reductions in manning resulting from the introduction of advanced technical change in manufacturing industry

	Total	Number employed at establishment	
		Fewer than 500	500 or more
Number initially employed in section	50	(26)[a]	(109)
Percentage reduction following change	*44*	*(42)*	*(47)*
Base: works managers reporting reductions in manning following the introduction of advanced technical change			
Unweighted	*59*	*24*	*35*
Weighted	*9*	*6*	*3*

[a] See note B.

sults. Reductions in manning were more common in larger plants. The scale of reductions appeared to be greater than the scale of any increases. The size of reductions was also greater in larger plants. On the other hand, it remained the case that both managers and manual stewards reported reductions in only about one quarter of cases and the number of people employed in sections affected by the changes being described represented a small proportion of total complements.

Staffing levels and advanced technical change in the office

Managers' accounts of the impact of the introduction of advanced technology upon the number of people employed in sections affected by office changes were very similar to their separate accounts of the effects on manning of shop floor innovations (see Table IX.6). As with applications on the shop floor, there were reports of some change in the number employed in affected sections in about 30 per cent of cases. Where change was reported, managers said that numbers were increased in about one third of the cases and decreased in about two thirds. As was often the case in our analysis, however, there were more signs of contrast between the

Table IX.6 The impact of advanced technical change on manning levels in manual sectors compared with its impact in the office

Column percentages

	All relevant establishments		Establishments where advanced technical change affected both categories of employee	
	Manual workers	Office workers	Manual workers	Office workers
Manning level in affected section(s)				
Increased	11	10	14	11
Decreased	19	18	29	26
No change	70	70	57	62
Not stated	*	2	*	1
Base: see footnote[a]				
Unweighted	*458*	*977*	*326*	*326*
Weighted	*212*	*500*	*78*	*78*

[a] See note to Table V.5.

implications of change for office workers compared with manual workers, when we confined analysis to establishments where separate changes were introduced affecting both categories (see the second half of Table IX.6). In these places, the introduction of advanced technology was more likely to influence the numbers employed among both categories but especially for manual workers. There was no sign, however, that the balance between increases and decreases was very different in these workplaces.

The general pattern of variation in the extent to which technological innovation in offices led to changes in staffing levels showed that this was one of the rare items which had a strong independent association with the total size of the enterprise of which the establishment was part. The more people employed by the enterprise the more likely was the size of affected sections to be decreased and the less likely was it to be increased (see Table IX.7). So far as the number of people employed at the establishment was concerned, it was among the establishments of intermediate size in white-collar terms that the size of sections was most frequently subject to changes, both favourable and unfavourable, following the introduction of computers and word processors (see Table IX.8).

Table IX.7 Proportion of workplaces where staffing was reduced by word processors/computers in offices affected

	Number of non-manual workers at establishment		
	25-49	50-149	150 or more
Total number of employees in organisation			
25-499	16	13	§[a]
500-9999	15	21	22
10000 or more	22	17	26

[a] Unweighted base too low for analysis.

There were substantial differences in the extent to which technological innovation in offices led to changes in staffing, depending upon whether establishments were independent or parts of a group and also, in the case of groups, whether they were overseas or domestically owned. Independent workplaces were the one category within which advanced technical change was more likely to lead to increases in staffing than to reductions. Innovations in establishments belonging to foreign-owned organisations

Table IX.8 Impact of word processors or computers on staffing in section or sections affected

Column percentages[a]

	Total	Number of non-manual workers at establishment 25-49	50-99	100-199	200-499	500-999	1000 or more
Manning level in affected section(s)							
Increased	10	11	8	13	10	4	3
Decreased	18	15	16	19	29	23	16
No change	70	70	76	67	59	72	80
Not stated	2	4	1	1	2	1	1
Base: establishments with 25 or more non-manual workers and experiencing change involving word processors or computers							
Unweighted	*977*	*129*	*165*	*220*	*223*	*137*	*103*
Weighted	*500*	*201*	*137*	*91*	*47*	*14*	*10*

[a] See note C.

Table IX.9 Proportion of workplaces where staffing in offices affected by change involving word processors or computers was reduced in relation to non-manual trade union recognition

	Number of non-manual workers	
	50-149	150 or more
Non-manual union recognised	23	34
Union not recognised	15	20

Base: private sector establishments with 25 or more non-manual workers and experiencing change involving word processors or computers

Table IX.10 Comparison of managers' and non-manual stewards' accounts of the impact of advanced technical change on staffing in section(s) affected

Column percentages

	Non-manual union recognised		Workplaces where both managers and stewards were interviewed about an advanced technical change	
	All managers	All shop stewards	Managers	Stewards
Manning levels in affected section(s)				
Increased	7	11	6	12
Decreased	18	17	23	21
No change	75	72	71	67
Base: as column heads[a]				
Unweighted	*747*	*520*	*437*	*437*
Weighted	*286*	*201*	*138*	*138*

[a] See note to Table VI.7.

were substantially more likely to lead to changes in staffing than the introduction of new technology in domestic establishments. Where there was change, it was also more likely to take the form of a reduction in staffing. The patterns in relation to the ownership of establishments were especially marked among smaller workplaces.

So far as industrial relations institutions were concerned, there was a marked tendency for frequencies of reductions and increases in staffing to be much more balanced in places where non-manual unions were not recognised. The chief difference, as Table IX.9 shows, was that increases in staffing following upon technological innovation were substantially less common where unions were recognised.

Non-manual stewards' accounts of the effect of introducing word processors and computers on staffing in offices
As Table IX.10 shows, the reports of stewards were broadly in line with those of managers in their respective accounts of the short-term impact of advanced technical change in offices upon manning. In contrast with the comparable table relating to manual workers, however (see Table IX.1), non-manual stewards were a little more likely than managers to report that the change led to an increase in staffing in the affected section or sections.

Change in the size of workforces over the previous four years

In addition to our information on the impact of particular technological changes upon the sections where they were introduced, we also had a measure of how the total number of people employed at each workplace had altered over the previous four years. This was derived from figures recorded on a basic workforce data sheet which we asked establishments to complete from their records before the interview was carried out. On the sheet was recorded the total number of people employed at the establishment, first, at the time it was completed and, secondly, four years previously. From these two figures we calculated the nature and extent of any change in the size of the workforce over the previous four years. Here we emphasise that the percentage changes referred to the total workforce, including both full-time and part-time employees and both manual and non-manual employees. Although our information on the size and composition of the workforce at the time of interview distinguished between the different sub-groups, it was not possible for us also to collect comparable figures for the previous period. Accordingly, we were not able to plot movements in the different categories of employee in relation to different forms of change.

Our information on the number of people employed at workplaces at different points in time represented a new and valuable source of data about change in employment. Two qualifications need to be made, however, about the use we make of the data in this chapter. First, as might have been expected, we had an unusually high level of non-response to the question asking for the number of employees four years previously. This meant that we had sufficient information to calculate percentage changes for only about four fifths of our workplaces. Secondly, the measure of change that we use in this chapter was one calculated to enable us to explore the pattern of variation as between different categories of workplace. Our average change represents an average derived from calculating the percentage change for each establishment and then calculating the average of those different percentages. It does not represent the aggregate of employees at all our establishments in 1984 expressed as a proportion of all employees in 1980(6). For this reason, among others, the overall change suggested in Table IX.11 is misleading because it fails to take account of the way in which there was a marked tendency for larger workplaces to have been more likely to reduce their complements. When percentages were weighted to reflect the number of people employed at establishments then there was clear evidence of an overall decline in employment. Nevertheless, despite any general limitations, the measure of change in employment at workplaces over a four year period that we use in this chapter represents a good basis for comparing the experiences of different kinds of workplace and adds substantially to previous information on the employment effects of new technology.

The first striking finding to emerge from the analysis of the number of people employed over the four year period was that the very large majority of establishments experienced substantial change (91 per cent). Fifty-three per cent had suffered a fall in the number of people employed, while 38 per cent enjoyed an increase. In those places where there was an increase, the scale of that rise was larger than the extent of the fall in places where there was a decline in the number of people employed.

There were major differences in the extent of change in relation to both size and sector. The variations in relation to size were especially strong, as already indicated. Overall, it appeared that larger workplaces were substantially more likely than smaller ones to experience reductions in staffing. For instance, workplaces employing 1,000 or more people reduced their workforces by 10 per cent, on average, while those employing fewer than 100 experienced a modest increase.[7] More detailed analysis, however, showed that this overall association between size and the extent of decline in numbers was almost wholly a consequence of the very strong association between the number of manual workers employed and the extent of decline. The strength and consistency of this association is shown in the first half of Table IX.11. The corresponding analysis in relation to the number of non-manual workers shows that any association was weak and inconsistent. In order to explore these relationships further, we looked at the association between falling employment and numbers of manual workers employed, holding the size of the non-manual workforce constant, and vice versa. Table IX.12 shows the resultant analysis. It is apparent that, when the number of non-manual employees was held constant, there was a marked tendency for workforces to have been more frequently reduced the larger the size of the manual workforce. But there was no corresponding tendency for the size of the

Table IX.11 Change in size of total workforce over previous four years in relation to (a) number of manual workers and (b) number of non-manual workers

Column percentages[a]

		Number of manual workers employed						
Change in workforce over previous four years	Total	None	1-24	25-49	50-99	100-199	200-499	500 or more
Increased	38	49	42	32	39	30	12	18
Remained similar	9	6	8	14	7	6	8	11
Decreased	53	44	50	54	54	64	71	73
Average change (per cent)	-	+7	+1	−2	+2	−6	−10	−14
Unweighted	*1597*	*133*	*348*	*270*	*198*	*199*	*232*	*217*
Weighted	*1588*	*206*	*615*	*419*	*191*	*90*	*50*	*18*

		Number of non-manual workers employed					
	Total	1-24	25-49	50-99	100-199	200-499	500 or more
Increased	38	35	43	38	46	40	24
Remained similar	9	9	9	10	7	14	16
Decreased	53	57	48	52	47	47	60
Average change (per cent)	-	−2	+3	+1	−1	+3	−6
Base: establishments where figures on number employed in 1980 were available							
Unweighted	*1597*	*360*	*263*	*271*	*237*	*239*	*222*
Weighted	*1588*	*747*	*431*	*231*	*94*	*52*	*23*

[a] See note C.

Table IX.12 Change in size of total workforce in relation to number of manual and non-manual workers employed

Percentage change

	Number of manual workers employed		
Number of non-manual workers employed	1-49	50-199	200 or more
1-49	−1	+2	−11
50-149	+3	−7	−10
150 or more	+3	−9	−12

non-manual workforce to have an independent association with the change in the size of the workforce.

So far as sector was concerned, private manufacturing establishments tended to have experienced a decline in jobs, while private services enjoyed a slight increase. In the public sector, nationalised industries suffered a loss of jobs, while public services enjoyed a very slight increase. This pattern is consistent with the tendency for job losses to be concentrated in places where relatively large numbers of manual workers were employed and with the general pattern of gross changes in employment as measured by official statistics. But sector and size, as measured by number of manual workers, remained independent sources of variation. Within manufacturing, for instance, it remained the case that workplaces were substantially more likely to have a smaller workforce the more manual workers were employed. Similarly workplaces in private services were most likely to have increased their complement within each size band (see Table IX.13).

Table IX.13 Change in size of total workforce in relation to sector and number of manual workers

Percentage change

	Private manufacturing	Private services	Public services
All establishments	−4	+3	+1
Establishments with 1-49 manual workers	−5	+2	+1
Establishments with 50-199 manual workers	−1	+2	−3
Establishments with 200 or more manual workers	−13	−2	−4

Table IX.14 Change in size of workforce over previous four years in relation to trade union recognition

Percentage change

	Manual union recognised	Manual union not recognised	Non-manual union recognised	Non-manual union not recognised
All establishments	−5	+7	−2	+4
Private manufacturing establishments	−13	+6	−15	–[a]
Private service establishments	−4	+6	−2	+5
Medium sized establishments[b]	−6	+16	−2	+9
Private manufacturing	−10	+23	−15	–
Private services	−5	+10	–	+11

[a] No change, on average.

[b] The analysis in the second half of the table is confined to workplaces of medium size, employing 50 to 199 manual workers or 50 to 199 non-manual workers, for the respective analyses.

So far as trade union organisation was concerned, there was a strong and consistent tendency for a decline in the size of workforces to be associated with trade union recognition and for an increase to be associated with lack of recognition. This contrast was especially marked in relation to arrangements regarding manual workers. The contrast persisted across size bands and sectors, as shown in Table IX.14. Moreover, when differences in size were taken into account, there was a marked tendency, especially in manufacturing industry, for a decline in numbers to be associated with trade union density among manual workers. The higher the density, the more the workforce decreased over the previous four years. Two observations may readily be made about the very strong association between trade union organisation, especially manual union organisation, and reductions in the size of workforces over time. The first is that, whatever the reasons for the association and whatever causal mechanisms may have contributed to it, the association between trade union representation and membership and decline was remarkably strong. The trend clearly has major implications for thinking about the future of the unions. Secondly, the association was not consistent with any simple notion that trade unions act as an obstacle to the run-down of workforces.

Change in the size of workforces in relation to the introduction of advanced technology

Having identified the overall pattern regarding changes in the size of the workforce, we have established a framework within which to explore any association between the introduction of advanced technology and job loss or gain, conscious of the main sources of variation that we need to take into account in evaluating any such association. For purposes of analysing the relationship between the use of new technology and changes in the size of the workforce over a four-year period, we adopted a number of measures of the use of microelectronics. First, we distinguished between places where manual workers operated the new technology and those where they did not. Secondly, we contrasted places that used computers and word processors and those that did not. Thirdly, we devised a variable based upon combinations of usage affecting manual workers and usage affecting office workers. Fourthly, we used our distinction between workplaces experiencing advanced technical change, conventional technical change, organisational change and no change, about which we had information for both manual workers and office workers. The results of the analysis were perhaps best summed up by the comparisons between workplaces that used applications of the new technology which affected both manual and non-manual workers and those that had applications affecting only one category, and also those where there was no application. We conclude this section with that comparison. We lead up to that conclusion by briefly reporting the results arising from analysis in relation to the other measures of the use of new technology.

The results of our initial, simple analysis suggested that there was some overall association between job loss and the application of advanced technology to the jobs of manual workers. In contrast, any association between the introduction of computers and word processors into office operations was, if anything, in the opposite direction. Table IX.15 shows that places where microelectronic technology was being used had, on average, experienced a three per cent fall in the size of their workforces, while there was no change, on average, among places where no microelectronic equipment was in operation. Perhaps more strikingly, the table also shows that the earlier advanced technology affecting manual workers was introduced, the larger was the overall job loss in the previous four years. In contrast, Table IX.16 shows that workplaces where there were computers or word processors had, on average, experienced a very slight increase in their workforces. Those which had no computers or word processors had suffered a very slight decline. In further contrast, there was no clear association between the time that computers and word processors were first introduced and any movements in the size of complements. Indeed, those first in the field appeared to have experienced the largest increase.

Secondly, we looked at movements in the complement of establishments in relation to whether they had introduced different forms of change in the previous three years. The pattern which emerged was consistent with the picture just described regarding the general use of microelectronics. As Table IX.17 shows, workplaces that introduced advanced technology affecting manual workers had, on average, suffered a reduction of five per cent in their workforce. The introduction of conventional technology affecting manual workers was associated with a substantially more modest reduction in complement. In cases where there had been no technologically-based change, the size of workforces tended, on average, to have remained much the same. Again, however, there was a different pattern arising from office applications of new technology. In places that introduced computers or word processors, the size of workforces had, on average, grown, while substantial reductions in the size of workforces were associated with the introduction of conventional new technology or organisational change. The pattern supported the indications from the initial analysis that manual applications of new technology were associated with job loss, while office applications were associated with the growth of jobs.

This, then, was the overall pattern, but it left open the possibility that the associations established were simply an incidental consequence of the fact that establishments in certain sectors and of a certain size were more likely, for instance, both to have introduced advanced technology affect-

Table IX.15 Change in size of total workforce in relation to the use of microelectronics affecting the jobs of manual workers

Percentage change

Change in workforce over previous four years	Not using micro-electronics	Using micro-electronics	Year of initial application: Before 1975	1975-1978	1979-1980	1981-1982	1983 or later
Increased	38	31	14	22	23	41	38
Remained similar	10	7	15	1	9	1	10
Decreased	52	62	72	77	68	57	52
Average change (per cent)	–[a]	–3	–13	–7	–8	+1	–
Base: establishments employing manual workers and providing workforce figures for both 1980 and 1984							
Unweighted	*999*	*465*	*54*	*87*	*121*	*97*	*87*
Weighted	*1133*	*249*	*12*	*43*	*54*	*53*	*85*

[a] No change, on average.

Table IX.16 Change in size of total workforce in relation to the use of computers or word processors affecting office staff

Percentage change

	Not using word processors or computers	Using word processors or computers	Year of initial application: Before 1975	1975-1978	1979-1980	1981-1982	1983 or later
Change in workforce over previous four years							
Increased	33	41	55	43	34	48	49
Remained similar	11	8	3	8	4	3	8
Decreased	56	51	42	48	62	48	44
Average change (per cent)	−1	+1	+18	+1	−9	+1	+6
Base: establishments providing workforce figures for both 1980 and 1984							
Unweighted	*312*	*1280*	*69*	*138*	*96*	*160*	*222*
Weighted	*561*	*1016*	*17*	*54*	*48*	*77*	*186*

Table IX.17 Change in size of total workforce in relation to recent technical and organisational changes

Percentage change

	Forms of change affecting manual workers				Forms of change affecting office workers			
	Advanced technical change	Conventional technical change	Organisational change	No changes	Advanced technical change	Conventional technical change	Organisational change	No changes
Change in workforce over previous four years								
Increased	31	36	38	38	44	22	32	38
Remained similar	9	10	12	9	7	13	7	10
Decreased	60	54	49	53	49	65	62	52
Average change (per cent)	−5	−1	–[a]	–	+2	−6	−4	–
Base: establishments providing workforce figures for both 1980 and 1984								
Unweighted	*480*	*205*	*124*	*735*	*871*	*88*	*81*	*552*
Weighted	*194*	*164*	*138*	*886*	*555*	*124*	*108*	*790*

[a] No change, on average.

ing manual workers and to have been subject to a contracting total workforce. To sort out how far that was the case, we carried out the analysis separately for different sectors and for different size bands within sectors.

As Table IX.18 shows, it was within private services that the general pattern we identified was most pronounced. Manual applications were associated with a much more marked decline in total complement, while office applications were associated with an increased workforce size, on average. In manufacturing industry, by way of contrast, there was a greater tendency towards uniform decline, regardless of whether advanced technology had been introduced on the shop floor or in offices. We have noted consistently throughout our report that the public sector generally lagged behind both private manufacturing and private services in its use of microelectronics. Applications affecting manual workers were especially rare, even in those public service establishments that employed manual workers. It is apparent, however, that those few public service establishments that did introduce microelectronic applications affecting manual workers constituted a deviant case so far as the general pattern was concerned. These establishments experienced an above-average increase in the total size of their workforces. The pattern in relation to computer or word processor applications to office operations, however, was more similar to the pattern for private services.

We also explored the extent to which the overall picture regarding both manual and non-manual microelectronic applications and the picture for particular sectors was associated with variations in the size of establishments. We established that the findings shown in Tables IX.16, IX.17 and IX.18 remained much the same when we compared the experiences of establishments of similar size.

As mentioned at the beginning of this section, our findings on the associations between the use of different forms of new technology and changes in the size of workforces were most simply summed up when we analysed changes in employment by combinations of usage affecting manual and non-manual workers, as in Table IX.19. The pattern of variation was most marked in relation to private services. Where manual workers were using the new technology, numbers had fallen and the decline was especially marked where word processors were also in use. In contrast, places using word processors but with no manual workers operating new technology were employing more people, on average.

The experience of independent establishments

One of the limitations of exploring the employment effects of new technology through the experiences of our workplaces was that they were not self-contained. Generally, they were parts of larger organisations. This was the case for 80 per cent of our establishments. So far as such work-

Table IX.18 Change in size of workforce over previous four years in relation to technical change in different sectors

	Manual workers		Forms of change			
	Using microelectronics	Not using microelectronics	Advanced technical change	Conventional technical change	Organisational change	No changes
Private manufacturing	−3	−5	−7	–[a]	(−9)[b]	−4
Unweighted	*304*	*209*	*262*	*98*	*34*	*134*
Private services	−5	+3	(−2)	(+6)	(+9)	+3
Unweighted	*57*	*331*	*41*	*33*	*26*	*360*
Public services	(+5)	+1	(+6)	(−4)	(−3)	+2
Unweighted	*38*	*393*	*38*	*49*	*47*	*297*
Independent establishments	(−1)	+7	(+13)	(+7)	§[c]	+3
Unweighted	*35*	*126*	*28*	*29*		*106*
	Office workers		**Forms of change**			
	Using computers or word processors	Not using computers or word processors	Advanced technical change	Conventional technical change	Organisational change	No changes
Private manufacturing	−4	−4	−2	(−19)	(−8)	−4
Unweighted	*473*	*55*	*349*	*20*	*25*	*135*
Private services	+4	+1	+4	(−)	§	+3
Unweighted	*354*	*106*	*233*	*25*		*184*
Public services	+4	−3	+7	(−4)	(−3)	–
Unweighted	*316*	*115*	*183*	*34*	*32*	*182*
Independent establishments	+7	(+2)	+6	§	§	+3
Unweighted	*130*	*40*	*92*			*65*

[a] No change, on average.

Table IX.19 Change in numbers employed between 1980 and 1984 in relation to the use of new technology affecting manual workers and in the office

Percentage change

	Using both word processors and microelectronics affecting manual workers	Using microelectronics affecting manual workers but no word processors	Using word processors but no microelectronics affecting manual workers	Neither using word processors nor microelectronics affecting manual workers
All establishments	−4	−3	+2	−1
Unweighted base[a]	*279*	*184*	*368*	*628*
Sectoral variations				
Private services	(−21)[b]	(−3)	+6	+2
Unweighted base	*26*	*31*	*114*	*221*
Private manufacturing	−4	−2	−9	−5
Unweighted base	*194*	*110*	*73*	*141*
Nationalised industries	(−11)	(−4)	(−5)	(−2)
Unweighted base	*24*	*28*	*40*	*49*
Public services	+7	§[c]	+4	−1
Unweighted base	*27*		*137*	*213*
Variations in relation to size				
25-99 employees	§	-	+2	-
Unweighted base		*57*	*74*	*350*
100-499 employees	−7	−10	+1	−3
Unweighted base	*57*	*89*	*149*	*215*
500-999 employees	−15	(−1)	+1	(−3)
Unweighted base	*82*	*25*	*77*	*45*
1000 or more employees	−16	§	−5	§
Unweighted base	*124*		*68*	

[a] Establishments employing both manual and non-manual employees and providing workforce figures for both 1980 and 1984.
[b] See note B.
[c] Unweighted base too low for analysis.

places were concerned, it was clear from our analysis of levels of decision-making in Chapter V that decisions about both the introduction of advanced technology and the size of workforces were often taken at higher levels in the organisation, for instance branch, divisional or head office level. In consequence, the employment effect of a decision to invest in new technology at one establishment might take the form of a decision to contract the size of the workforce at a second.

Accordingly, it seemed appropriate further to assess the implications for jobs of introducing new technology through our data by focusing upon independent establishments where the consequences for employment, either reductions or expansions, could not so easily be absorbed by other establishments. Unfortunately there were not sufficient independent establishments in our sample to carry out that analysis separately for different sectors and size bands. The third line in Table IX.18, however, showed the analysis for independent establishments. Once again, the pattern revealed that manual applications were associated with job loss, while office applications were associated with job growth. The majority of independent establishments, however, were in the private service sector where this pattern was generally most pronounced.

Patterns of Workforce Reduction and Recruitment

Further opportunities to explore any association between employment and advanced technological change arose from our questions in the interview schedule about, first, manpower reductions and, secondly, recruitment over the previous year. Overall, a very substantial minority (41 per cent) of our establishments reported that they had had reason to reduce the size of their workforce in the previous year. Rarely, however, did they report that the reason for the reduction was the introduction of new technology. Reorganisations, lack of demand for goods or services, the imposition of cash limits, and the introduction of cost reduction exercises were all blamed for manpower reductions very much more frequently than the introduction of new machines or equipment. Only four per cent of all establishments reported that there had been reductions in one or more sections of their workforce in the previous year as a result of the introduction of new technology. Even in those establishments where we knew that new technology had recently been introduced, this proportion increased only slightly. Table IX.20 shows that those establishments with advanced technology affecting manual workers were very slightly more likely than average to have experienced manpower reductions in the previous year. In relation to experience of our three distinct forms of change, however, those which introduced organisational change only were more likely to experience recent reductions in their workforces than those which introduced new advanced technology. But places that had

Table IX.20 Extent and causes of manpower/staff reductions in previous 12 months in relation to the use of microelectronics affecting manual jobs

Column percentages[a]

	All establish-ments	Using micro-electronics	Not using micro-electronics	Forms of change in previous three years			
				Advanced technical change	Conventional technical change	Organisa-tional change	No changes
Manpower/staff reduction in previous year	41	43	40	41	43	55	38
No manpower/staff reduction in previous year	59	57	60	59	57	45	61
Reasons for reduction[b]							
Reorganisation	15	19	13	14	19	26	12
Lack of demand	14	12	15	14	15	20	14
Cash limits	12	11	12	10	8	18	11
Cost reduction	10	7	9	11	8	20	9
New technology	4	9	3	10	4	1	3
Other reasons	3	2	3	3	3	2	4
Base: all establishments employing manual workers							
Unweighted	*1853*	*547*	*1306*	*478*	*258*	*173*	*944*
Weighted	*1749*	*311*	*1438*	*254*	*214*	*178*	*1102*

[a] See note C.
[b] See note F.

introduced advanced technology in the previous three years were more likely than those introducing conventional technology to attribute their recent manpower reductions to new technology. One in every ten workplaces experiencing advanced technical change gave the introduction of new plant or equipment as the reason for reductions.

The association between the use and introduction of computers or word processors in the office and recent experience of workforce reductions (see Table IX.21) was very similar to that between manual applications and workforce reductions. The rate of reductions was close to the norm in places where computers or word processors had been introduced. Places where conventional technical change or organisational change had been introduced were more likely to have experienced reductions. Under all circumstances, new technology remained the least common of the major reasons for manpower reductions.

The pattern of association between both manual and non-manual applications of new technology and manpower reductions, shown in Tables IX.20 and IX.21, remained constant when establishments of similar size were compared. In relation to sector, the use of microelectronics affecting manual workers was most likely to be associated with manpower reductions in private services. This was consistent with the association we found earlier between declining employment over the previous four years and the use of advanced technology affecting manual workers. In both private manufacturing and public services, by way of contrast, establishments operating advanced technology affecting manual workers were, if anything, very slightly less likely to have experienced recent manpower reductions. So far as the introduction of computers and word processors were concerned, any association with recent reductions in staffing was very similar across all sectors.

As well as looking at the extent to which establishments reduced their workforces in the recent period, we also explored the methods they used for any reduction. Our results confirmed earlier research, showing that natural wastage, the redeployment of staff within the establishment, and early retirement were the most common methods used by British employers to reduce the size of their workforces(8). They resorted to redundancy substantially less frequently and by inference, only when these less painful methods had failed. Even when redundancy became unavoidable, the nominally voluntary method was used with similar frequency to enforced redundancy. In consequence, although 41 per cent of workplaces introduced manpower reductions in the previous year, only one in ten imposed any enforced redundancies. The indications from our findings were that such redundancies were very marginally more common in places that employed advanced technology than in those that did not.

Table IX.21 Extent and causes of manpower/staff reductions in previous 12 months in relation to the use of computers or word processors

Column percentages[a]

	All establishments	Using computers or word processors	Not using computers or word processors	Forms of change in previous three years: Advanced technical change	Conventional technical change	Organisational change	No changes
Manpower/staff reduction in previous year	41	44	36	42	51	46	39
No manpower/staff reduction in previous year	59	56	64	58	49	54	61
Reasons for reduction[b]							
Reorganisation	15	18	10	18	21	25	10
Lack of demand	14	16	12	15	11	14	15
Cash limits	12	13	10	10	19	16	11
Cost reduction	10	12	7	11	12	17	8
New technology	4	6	2	9	4	2	2
Other reasons	3	3	4	3	6	2	3
Base: all establishments employing non-manual workers							
Unweighted	*2010*	*1587*	*423*	*1075*	*106*	*109*	*720*
Weighted	*1985*	*1248*	*737*	*693*	*144*	*134*	*1014*

[a] See note C.
[b] See note F.

Methods of adjustment to the need for reductions in manning in manufacturing industry

Our general information about the ways in which managements brought about any reductions in workforces that they judged to be necessary was supplemented by reports from our interviews with works managers about how they achieved particular reductions required by specific technical or organisational changes. We showed earlier in this chapter that the introduction of advanced technical change affecting manual workers tended to lead directly to reductions in manning in about one case in every five. Our interviews with works managers in manufacturing industry gave us the opportunity to ask how these particular reductions were brought about. Their answers revealed that there were clear differences between the methods used by managements to cut numbers, when faced with the general need to reduce the overall size of their workforces, and the methods used to cut the number of people working in a particular section, when required to do so by the introduction of new technology. First, when new machines were being introduced, the range of methods used was very much more restricted. Secondly, the dominant method was internal redeployment. Thirdly, all the other main methods used generally, including attrition or wastage, early retirement, voluntary redundancy, and especially enforced redundancy, were reported very much less frequently. As might be expected, when the focus was a reduction in the size of an identifiable sub-group at the workplace required by the introduction of a particular change, then, in practice, a more limited range of options was open to managements than when the focus was the need to reduce their overall complement. The implication of our findings was that, in some cases, there was a sequential process of reduction. Initially, managements used redeployment to reduce the size of sections needing fewer people owing to technical innovation. Subsequently, when they judged they were employing too many people overall, they deployed the full battery of modern manpower practices to reduce the size of their total workforce.

Recruitment in the Previous Year

While nearly one half of all workplaces introduced manpower reductions in the previous year, it was also true that the large majority engaged in at least some recruitment over the same period. Tables IX.22 and IX.23 show that places using the new technology affecting manual workers were very slightly less likely to recruit than others, but where computers or word processors were being used, recruitment was slightly more common than where they were not. The biggest contrast, however, between places where the new technology was used and those where it was not lay in the frequency with which different categories were recruited. Where workplaces used advanced technology affecting manual workers, they

Table IX.22 Extent and nature of recruitment in previous 12 months in relation to the use of microelectronics affecting manual jobs

Column percentages

	All establishments	Using micro-electronics	Not using micro-electronics	Forms of change in previous three years: Advanced technical change	Conventional technical change	Organisational change	No changes
Some recruitment in previous year	85	83	86	80	85	86	87
No recruitment in previous year	15	17	14	20	15	14	13
Categories recruited[a]							
Manual – unskilled	34	40	39	42	46	42	37
– semi-skilled	23	34	25	32	35	30	24
– skilled	22	39	23	35	32	19	23
Non-manual							
– clerical/admin.	43	40	39	39	40	30	40
– supervisory	10	11	11	12	9	19	9
– junior technical/prof.	28	29	27	28	27	32	26
– senior technical/prof.	18	17	17	19	11	25	17
– middle/senior management	18	22	18	27	19	23	16
Base: all establishments employing manual workers							
Unweighted	*1853*	*547*	*1306*	*478*	*258*	*173*	*944*
Weighted	*1749*	*311*	*1438*	*254*	*214*	*178*	*1102*

[a] See note F.

Table IX.23 Extent and nature of recruitment in previous 12 months in relation to the use of computers or word processors

Column percentages[a]

	All establishments	Using computers or word processors	Not using computers or word processors	Forms of change in previous three years: Advanced technical change	Conventional technical change	Organisation change	No changes
Some recruitment in previous year	85	89	80	91	88	73	83
No recruitment in previous year	15	11	20	9	12	27	16
Categories recruited[b]							
Manual – unskilled	34	34	36	34	20	34	37
– semi-skilled	23	24	22	24	19	21	24
– skilled	22	23	21	25	17	15	22
Non-manual:							
– clerical/admin.	43	55	24	63	52	30	30
– supervisory	10	12	6	13	7	7	8
– junior technical/prof.	28	35	18	36	30	26	23
– senior technical/prof.	18	22	11	27	9	17	14
– middle/senior management	18	24	10	29	12	8	14
Base: all establishments employing manual workers							
Unweighted	*2010*	*1587*	*423*	*1075*	*106*	*109*	*720*
Weighted	*1985*	*1248*	*737*	*693*	*144*	*134*	*1014*

[a] See note C.
[b] See note F.

were much more likely to recruit skilled and semi-skilled manual workers. They recruited other categories in our list with similar frequency to places that did not operate new technology. Places that used computers or word processors were twice as likely as others to have taken on secretarial, clerical or administrative staff, which was the most common area of recruitment. They were also twice as likely to have taken on supervisory, technical, professional and management grades. It was clear that the use of advanced technology had a major impact upon the categories of people employed. The analysis reported earlier in the chapter suggested that it also had a modest impact upon the number of people employed. In combination, the results highlight the way in which advanced technical change has led to an acceleration in the trend in the pattern of employment over the past quarter of a century.

Footnotes

(1) C. Jenkins and B. Sherman, *The Collapse of Work*, Eyre Methuen, London, 1979.

(2) Jim Northcott, Michael Fogarty and Malcolm Trevor, *Chips and Jobs: Acceptance of New Technology at Work*, PSI, No.648, 1985.

(3) Northcott's 1983 PSI survey of manufacturing showed a modest net loss in jobs as a result of microelectronic technology. But, perhaps more instructively, his most recent, 1985 survey suggested that loss of jobs as a result of technical innovation was growing.

(4) For an estimate of the overall effect of the use of new technology upon jobs, including indirect effects, see R.M. Lindley *et al.*, *Review of the Economy and Employment*, University of Warwick, Institute of Employment Research, Spring, 1982.

(5) W.W. Daniel and Neil McIntosh, *The Right to Manage*, Macdonald, London, 1972.

(6) Further information on change in employment is provided by Neil Millward and Mark Stevens, *British Workplace Industrial Relations* 1980–1984, Gower 1986.

(7) The effect of our method of sample selection described in the Technical Appendix to our companion volume (Millward and Stevens, 1986, ibid.) will have enhanced the growth in the 25–49 size band.

(8) W.W. Daniel, *The Impact of Employment Protection Laws*, PSI, No.577, 1978.

X Technical Change and Increases in Earnings

The implications of advanced technology for levels of pay have surprisingly received comparatively little attention from commentators(1). The possible effects upon employment and job security have attracted a great deal of speculation and analysis. The impact upon the quality of jobs has been the focus of substantial debate over the years. But so far as the consequences for the pay of workers directly affected by the introduction of advanced technical change or employed by organisations that make use of advanced technology are concerned, it is difficult to find much evidence or comment. In view of the part that financial remuneration plays among the rewards that workers derive from their employment, this omission is remarkable. Job security is certainly important to workers, especially at a time of high unemployment, and they also value interesting, satisfying and congenial work. But it would be difficult to argue that the space devoted to employment and quality of work in all that has been written about new technology, compared with the space devoted to the implications for pay, fairly reflects the priorities of workers in their expectations and requirements of paid employment. A cynical explanation of the comparative lack of comment about levels of pay under advanced technology would be that it has suited neither the opponents nor the proponents of technical change to focus attention upon the issue. Critics of technical change have been reluctant to do so because, as we show in this chapter, the introduction of advanced technology has been associated with higher rates of pay and the pattern revealed would not support their case. On the other hand, the strongest promotors of change have frequently been employers or their representatives who may not have wanted to foster the idea that the introduction of technical change frequently carried a price tag.

In principle, there is good reason to expect that the introduction of new technology would tend to lead to higher earnings for the workers involved. New technology should lead to higher productivity, and there are a range of mechanisms in our system of pay determination through which increases in productivity are translated into pay increases for workers. These include payment by results on an individual, group, establishment or enterprise level; the *ability to pay* criterion and the *increased productivity* criterion in periodic bargaining over rates of pay; the negotiation of agreements on change, within which rates of pay are expressly increased as part of the arrangements for introducing change; and comparability claims from groups working alongside those who directly benefit from change. In addition, technical change frequently leads, as we saw in Chapter VII, to changes in the skill, responsibility and degree of flexibility involved in jobs, which may give rise to upgrading.

We are able to provide some measure of the extent to which increases in earnings do, in practice, result from technical change and to start to fill the gap in information about the implications of advanced technology for levels of pay, through two sets of questioning in our present survey. First, in cases where managers reported an advanced technical change in the previous three years, we asked whether the change led to an increase or a decrease in the pay of the workers directly affected. Where it led to an increase, we asked about the medium through which the increase was received. These questions were asked separately in relation to changes affecting both manual workers and office workers. They were also asked of both manual and non-manual union representatives, where appropriate. Secondly, at a quite different stage in the interview, we asked managers some detailed questions about the pay and hours of different categories employed at the establishment. This enabled us to explore the extent to which both gross earnings and hourly rates were associated with the extent to which workplaces were using advanced technology. Initially in this chapter, we look at accounts of the impact upon earnings of particular changes. In the second half of the chapter, we analyse the general pattern of earnings in relation to the use of advanced technology.

The Impact of Particular Technical Innovations Upon the Pay of the Manual Workers Directly Affected

The separate accounts of managers and shop stewards of the implications of the most recent major change they experienced for the earnings of the manual workers directly affected provided one example where the respective pictures provided by their answers were substantially different. Accordingly, it is worth our while in this instance to start with the comparison between the reports of managers and those of stewards (see Table X.1). It should be noted that managers' answers in the first column of the table are confined to cases where manual unions were recognised, but the pattern did not differ markedly from that for all cases where managers reported change (see Table X.2). It is apparent, especially when the comparison was confined to workplaces where we interviewed both managers and stewards about a similar change, that stewards were more likely than managers to report that the technical change led to a reduction in earnings and slightly less likely to report that it led to an increase. Of course, anyone familiar with the complexity of many systems of pay for manual workers and the scope they provide for confusion between basic

Table X.1 Comparison of managers' and manual stewards' accounts of the impact of advanced technical change on the earnings of the manual workers affected

Column percentages[a]

	Manual union recognised		Workplaces where both managers and stewards were interviewed about an advanced technical change	
	All managers	All shop stewards	Managers	Stewards
Implications of change for earnings				
Increased	24	25	34	28
Decreased	3	9	2	10
No change	73	65	63	62
Medium/source of any increase[b]				
Regrading	9	14	18	17
Bonus payments (PBR)	10	6	12	6
Change agreements (increased rates)	9	15	12	20
More overtime	5	6	5	7
Other answer	3	3	4	2
Base: as column heads[c]				
Unweighted	*405*	*288*	*205*	*205*
Weighted	*160*	*101*	*72*	*72*

[a] See note C.
[b] See note F.
[c] See note to Table VI.7.

rates, supplementary payments of different kinds, hourly rates and gross earnings, will not be surprised that it was perfectly possible for two people to describe quite differently the impact of a particular change upon pay. This was especially true when the two separate accounts were derived from people with different negotiating perspectives, which were likely to encourage them to focus upon aspects of the change that best suited their case. In such circumstances, it may be more surprising that the reports from managers and shop stewards were so similar. The suggestion that the reason for any differences there were between the two accounts lay primarily in the complexity of some systems of pay for manual workers was supported by the fact that, as we show later, there was much closer accord between the reports of managers and non-manual stewards when they were describing the effects of change upon the earnings of office workers.

Moreover, the two broad conclusions, that would be drawn separately from the accounts of manual stewards and managers, were similar. First, the technical change led to no change in earnings in the majority of cases. Secondly, where there was any change, it took the form of an increase in the large majority of instances. In addition, the media through which any increases were received appeared broadly similar from both sets of account. Stewards were more likely to attribute any increase to an agreement on change specifying higher rates, however, while managers were more likely to attribute any increase to increased earnings under some system of payment by results.

When we looked at sources of variation in the extent of change in pay, we found no consistent association between such changes and the size of workplaces, according to either account. But there was a substantial difference in relation to sector, as Table X.2 shows. It was very much more common for changes to result in increases in earnings in the private sector, especially in manufacturing industry. It might have been expected that this would be the case, as systems of payment by results (PBR) were more common in the private sector and technological change could be expected to have an impact upon such systems. But it is clear that it was not only through PBR that workers in manufacturing industry were more likely to achieve increases as a result of technical innovation. They were also very much more likely to have benefited from regrading, the negotiation of higher rates through a formal change agreement, and to a lesser extent through more overtime. It is notable that the proportion of cases where increases through change agreements were reported was very similar to the proportions where both managers and manual stewards reported in Chapter VI that the changes were negotiated with union representatives and subject to their agreement. Such agreements were the chief source of any increase for manual workers in public services, but those receiving increases in this sector were very much concentrated in workplaces where larger numbers of manual workers were employed.

There was, surprisingly, very little overall difference, according to management accounts, in the extent to which workers received increases in relation to whether or not manual unions were recognised (see Table X.3). Indeed, when analysis was confined to the only size band where there were sufficient numbers for comparison, it appeared that workers in that minority of cases where unions were not recognised were more likely to receive increases. There was, however, a slight tendency for increases in pay to be more commonly associated with change as union density rose, especially when comparisons were made between workplaces of similar size.

Table X.2 Impact of advanced technical change on earnings in relation to sector

Column percentages[a]

	All establishments	Private manufacturing	Private services	Nationalised industries	Public services
Implications of change for earnings					
Increased	24	35	18	6	8
Decreased	2	3	–	2	5
No change	73	62	82	92	87
Not stated	1	1	–	–	–
Medium/source of any increase[b]					
Regrading	10	17	1	1	1
Bonus payments (PBR)	9	11	13	1	2
Change agreements (increased rates)	8	11	2	3	6
More overtime	4	7	2	2	1
Other answer	4	5	4	*	–
Base: establishments with 25 or more manual workers and having advanced technical change					
Unweighted	*458*	*286*	*59*	*63*	*50*
Weighted	*212*	*116*	*44*	*26*	*28*

[a] See note C.
[b] See note F.

Advanced Technical Changes and the Earnings of Office Workers

As we mentioned earlier, there was not the same contrast between the reports of managers and non-manual stewards about the impact of recent technical changes upon the earnings of office workers as was apparent at the manual level. Indeed, as Table X.4 shows, it would have been difficult for the reports to have been closer, both in relation to the description of any impact upon earnings and with regard to the medium for any increase. It was also apparent that the broad picture revealed by reports of any impact upon the pay of office workers was similar to that for manual workers. There was no change in the majority of cases. Where there was change, it took the form of an increase in pay in the large majority of

Table X.3 Impact of advanced technical change on earnings in relation to union recognition

Column percentages

	All establishments		Medium size establishments with 50-199 manual workers	
	Manual union recognised	Union not recognised	Manual union recognised	Union not recognised
Implications of change for earnings				
Increased	24	24	24	(50)[a]
Decreased	3	–	3	–
No change	72	76	72	(50)
Not stated	1	–	1	–
Medium/source of any increase[b]				
Regrading	9	11	9	(12)
Bonus payments (PBR)	10	5	12	(13)
Change agreements (increased rates)	9	5	12	(14)
More overtime	5	3	10	(9)
Other answer	3	7	2	(19)
Base: establishments with 25 or more manual workers and experiencing advanced technical change				
Unweighted	*405*	*53*	*100*	*26*
Weighted	*160*	*52*	*64*	*18*

[a] See note B.
[b] See note F.

cases. Table X.5 compares the effects of recent technical changes upon the pay of office workers and manual workers according to the accounts of managers. The pictures were remarkably similar. This was certainly the case when we compared all office applications with all shop floor changes. There were more differences in the media for the increases. Office workers were substantially more likely to benefit from upgrading, while manual workers were especially more likely to enjoy higher earnings under some system of payment by results but also to benefit from higher rates negotiated as part of a change agreement and increased overtime working.

Table X.4 The impact of advanced technical change upon the earnings of the office workers affected

Column percentages

	Non-manual union recognised		Workplaces where both managers and stewards were both interviewed about an advanced technical change	
	All managers	All shop stewards	Managers	Stewards
Implications of change for earnings				
Increased	21	22	23	23
Decreased	*	1	*	1
No change	79	77	77	76
Medium/source of any increase				
Regrading	17	19	17	19
Bonus payments (PBR)	1	1	1	1
Change agreements (increased rates)	6	7	7	7
More overtime	*	1	1	1
Other answer	3	1	3	1
Base: as column heads[a]				
Unweighted	*747*	*520*	*437*	*437*
Weighted	*286*	*201*	*138*	*138*

[a] See note to Table VI.7.

When we focussed our analysis upon those workplaces which reported changes both in offices and on the shop floor, through which we could analyse the comparative experiences of office and manual workers in the same workplaces, there was a more marked contrast in the experience of the two groups (see second half of Table X.5). First, manual workers appeared more likely to enjoy higher earnings as a result of the change. Secondly, the contrasts between the media through which the respective groups received increases became more marked. Of course, the types of establishment that had tended to introduce technical change in the office and on the shop floor had distinctive characteristics. They were concentrated among the large workplaces and in the manufacturing sector.

So far as general sources of variation in the distribution of change in the earnings of non-manual workers following technological change were concerned, it appeared, on the face of it, that office workers were less likely to receive increases in the largest workplaces and in the public sector. More detailed analysis, however, showed, first, that when the number of non-manual workers employed was adopted as the measure of size, office workers in the intermediate category, employing 200 to 499 non-manual workers, were most likely to enjoy enhanced pay. Secondly, there appeared to be little difference between the experience of office workers in private manufacturing, private services or public services, when differences in size were taken into account. It remained the case, however, that office workers employed in nationalised industries were less likely than others to receive increased earnings. Similarly, it appeared, superficially, that office workers employed by foreign-owned companies were more likely to enjoy increases, as were those employed by independent

Table X.5 Managers' separate accounts of the impact of advanced technical changes upon the earnings of manual workers and office workers

Column percentages

	All relevant establishments		Establishments where advanced technical change affected both categories of employee	
	Manual workers	Office workers	Manual workers	Office workers
Implications of change for earnings				
Increased	24	24	29	23
Decreased	2	*	2	*
No change	73	74	68	76
Not stated	1	2	1	1
Medium/source of any increase[a]				
Regrading	10	18	10	20
Bonus payments (PBR)	9	1	14	*
Change agreements (increased rates)	8	5	8	3
More overtime	4	2	9	1
Other answer	4	4	4	1
Base: As column heads[b]				
Unweighted	*458*	*977*	*326*	*326*
Weighted	*212*	*500*	*78*	*78*

[a] See note F.
[b] See note to Table V.5.

establishments rather than places belonging to groups. More detailed analysis, however, within size bands revealed that the pattern was variable.

So far as industrial relations institutions were concerned, it certainly appeared that trade union representation did little to increase the chances that office workers had of increasing their earnings as a result of technological change. Indeed, as Table X.6 shows, office workers in places where non-manual unions were recognised tended to be less likely to receive increased earnings than counterparts in places where unions were not recognised. The pattern mirrored that which we identified earlier among manual workers, and remained similar when categories were compared within size bands. There were also modest differences in the media for pay increases as between places where unions were recognised and those where they were not. As might be expected, office workers in union shops were more likely to receive increases through negotiated change agreements, while counterparts in non-union shops were more likely to benefit from upgrading. There was no sign that non-manual union density was systematically associated with whether or not office workers received increases.

As was the case with technological change affecting manual workers, there were associations between the extent to which the change resulted in pay increases and other features of the change. Essentially, however, these patterns revealed that the more likely there was to have been a change in one dimension, the more likely there was to have been a change in another. For instance, if the level of manning or staffing was changed, office workers were substantially more likely to receive increases. This was the case whether manning was increased or decreased, though it was slightly more common where manning was increased. Similarly, office workers were substantially more likely to receive increases in pay where a change in the level of interest in the job was reported. Again, it made little difference if managers reported that job interest was enhanced or diminished as a result of the introduction of new equipment. In both instances, similar proportions received increases in pay. The implication of these patterns (see Table X.7) is that whether the pay of office workers directly affected by change varied in response to the innovation depended more upon *the extent* of the change than the *nature or direction* of the change.

Pay Levels in Relation to the Use of Advanced Technology

We have established that in the majority of cases the introduction of advanced technology did not result in any immediate change in earnings for the workers directly affected. In the substantial minority of cases where there was any change, it generally took the form of an increase. This was

Table X.6 Implications of introducing change involving computers and word processors on earnings in relation to union recognition

	All establishments		Small establishments with lower than 50 non-manual workers		Large establishments with 150 or more non-manual workers	
	Non-manual union recognised	Union not recognised	Non-manual union recognised	Union not recognised	Non-manual union recognised	Union not recognised
Implications of change for earnings						
Increased	21	27	17	31	25	29
Decreased	*	–	–	–	*	–
No change	77	70	79	64	74	71
Not stated	2	3	4	5	1	–
Medium/source of any increase[a]						
Regrading	16	20	14	20	19	25
Bonus payments (PBR)	1	–	–	–	1	–
Change agreements (increased rates)	6	2	3	3	8	*
More overtime	*	4	–	3	*	5
Other answer	3	6	3	7	3	3
Base: establishments with 25 or more non-manual workers and introducing computers or word processors						
Unweighted	*747*	*230*	*55*	*74*	*482*	*76*
Weighted	*286*	*213*	*83*	*119*	*84*	*23*

[a] See note F.

Table X.7 Implications of introducing change involving computers or word processors on earnings in relation to other aspects of the change

	Effect of change on staffing			Effect of change on job interest		
	Increased	No change	Decreased	More	No change	Less
Implications of change for earnings						
Increased	41	19	36	29	11	33
Decreased	–	*	*	*	*	–
No change	59	81	64	71	88	65
Not stated	–	*	–	–	1	1
Medium/source of any increase						
Regrading	36	13	27	23	5	28
Bonus payments (PBR)	–	*	2	1	*	–
Change agreements (increased rates)	3	4	6	7	1	4
More overtime	4	1	4	2	1	5
Other answer	8	3	6	3	5	5
Base: establishments with 25 or more non-manual workers and introducing computers or word processors						
Unweighted	*79*	*638*	*245*	*634*	*224*	*53*
Weighted	*50*	*351*	*88*	*302*	*133*	*27*

true for both office applications of advanced technology and for changes affecting manual workers, but more so at the office level. In general, the changes tended to lead directly to increases in pay in about one quarter of cases. We now go on to analyse any general associations there were between the extent to which workplaces used advanced technology and the earnings of different categories.

First, however, a word is required about the nature of our information on levels of pay. We sought to establish the gross weekly pay of a *typical* (defined, if necessary, as the *modal*) man or woman in each of five broad occupational categories: unskilled manual workers; semi-skilled manual workers; skilled manual workers; clerical, secretarial and administrative workers; and supervisors or foremen or women. Having established the pay level, we asked how many hours a week it was necessary to work in order to earn that amount. We were thus able to calculate average hourly rates of pay for typical workers in different grades. The chief complication relating to our information on pay, however, is that it was not possible, because of pressure on interview time, for us to ask about the earnings of both men and women separately in each grade at each workplace. Accordingly, we established initially, where establishments had any employees in a particular category (such as skilled manual workers), whether the majority were men or women. We then focussed the questions about pay on the majority gender. Thus, in relation to any particular category, at any particular establishment, we had information about earnings for male workers *or* female workers. Accordingly, when we analyse our information on levels of earnings for men, the establishments from which the information was derived will vary from grade to grade. This accounts for the different base numbers for each line in Table X.8. The same is true of our results on the pay of women.

Having explained the complications in our information about pay levels, the main results of, or analysis in relation to, the use of advanced technology were fairly straightforward. It did appear that the use of microelectronics was associated with higher levels of earnings. This was the case whether we focussed upon gross weekly pay or upon hourly rates. The association held for both manual and office applications and for the earnings of both men and women. We found the most interesting patterns of association between earnings and the use of new technology when we analysed levels of pay in relation to combinations of usage in the office and on the shop floor. We start the analysis, however, by looking, first, at variations in pay associated with whether or not manual workers operated processes using microelectronics. Secondly, we look at variations in relation to use of computers and word processors. The analysis in relation to combinations of usage then concludes the chapter.

Table X.8 Gross weekly earnings of different categories in relation to the use of microelectronics affecting the jobs of manual workers

£ per week

	Men		Women	
	Using micro-electronics	Not using micro-electronics	Using micro-electronics	Not using micro-electronics
Unskilled manual	110	104	70	52
Unweighted base[a]	*338*	*544*	*144*	*553*
Semi-skilled manual	124	117	98	90
Unweighted base	*297*	*374*	*125*	*358*
Skilled manual	160	147	124	109
Unweighted base	*324*	*437*	*133*	*372*
Clerical/secretarial	109	109	99	95
Unweighted base	*330*	*526*	*155*	*537*
Supervisory/foremen/women	174	164	132	135
Unweighted base	*326*	*480*	*141*	*428*

[a] In each instance the base for the calculation of average pay is establishments employing manual workers where the typical pay of men or women, whichever was appropriate, was provided by management respondents for the particular category.

Earnings and applications of microelectronics affecting manual workers
Table X.8 shows that the earnings of male manual workers were higher in workplaces which used advanced technology which affected manual workers. The contrast between places operating new technology and those that did not was especially marked so far as skilled manual workers were concerned. It is striking that, while there were differences between the earnings of all the manual grades and the first-line managers of manual workers, according to whether or not there were manual applications of new technology, there was no difference in the earnings of the one non-manual grade included. In cases where women were in the majority, the pattern was not dissimilar to that for men. Indeed, it looked as though the difference in the earnings of women between workplaces using advanced technology and those not doing so might be even greater, especially so far as unskilled and semi-skilled grades were concerned.

We repeated the analysis in Table X.8 within different sectors and within establishments of different size. The pattern revealed by Table X.8 tended to persist with remarkable consistency. So far as male earnings were concerned, within each size band, the pay of manual workers in places operating advanced technology was generally higher than that of counterparts in places where they did not. This was especially so in the case of skilled manual workers, with the exception of those employed by the largest workplaces (as measured by the number of manual workers employed). The differential enjoyed by women manual workers when they were employed in places using advanced technology also persisted throughout the size bands.

When we analysed differentials within the different broad sectors, however, more exceptional cases emerged. The patterns for both men and women in private manufacturing and for men in private services were very similar to the general pattern. Inexplicably, however, the earnings of women in private services showed the differential to be in favour of those working in places where there was no advanced technology affecting manual workers. The number of places within private services which operated advanced technology affecting manual workers was small, and this was especially so where women were in the majority among manual workers.

The most striking deviant case was provided by the nationalised industries where the earnings of both skilled and semi-skilled male manual workers were higher in places operating no advanced technology than they were in places that did have applications affecting manual workers. This may simply have been a consequence of the way in which micro-

Table X.9 Gross weekly earnings of different categories in relation to the use of word processors or computers

£ per week

	Men		Women	
	Some use of word processors or computers	No use	Some use of word processors or computers	No use
Unskilled manual	108	100	56	52
Unweighted base[a]	*726*	*154*	*523*	*175*
Semi-skilled manual	121	115	98	81
Unweighted base	*564*	*105*	*383*	*101*
Skilled manual	157	136	117	104
Unweighted base	*637*	*122*	*399*	*106*
Clerical/secretarial	110	104	98	90
Unweighted base	*716*	*140*	*535*	*158*
Supervisory/foremen/women	169	160	139	129
Unweighted base	*675*	*131*	*441*	*129*

[a] See note to Table X.8.

Table X.10 Gross weekly earnings of different categories in relation to the use of new technology affecting manual workers and in the office

£ per week

	Using both word processors and and micro-electronics affecting manual workers	Using micro-electronics affecting manual workers but no word processors	Using word processors but no microelectronics affecting manual workers	Neither using word processors nor microelectronics affecting manual workers
Women				
Unskilled manual	70	48	70	53
Unweighted base[a]	*80*	*198*	*64*	*354*
Semi-skilled manual	113	92	92	89
Unweighted base	*65*	*140*	*60*	*218*
Skilled manual	136	121	118	106
Unweighted base	*172*	*147*	*61*	*225*
Clerical/secretarial	105	102	96	93
Unweighted base	*81*	*202*	*74*	*335*
Supervisory/foremen/women	151	154	124	131
Unweighted base	*76*	*159*	*65*	*269*

Table X.10 Continued

Men				
Unskilled manual	110	116	109	101
Unweighted base	*204*	*204*	*133*	*339*
Semi-skilled manual	128	131	121	114
Unweighted base	*184*	*134*	*112*	*239*
Skilled manual	166	162	157	143
Unweighted base	*197*	*159*	*126*	*277*
Clerical/secretarial	113	128	106	103
Unweighted base	*200*	*202*	*130*	*324*
Supervisory/foreman/women	185	169	169	162
Unweighted base	*198*	*176*	*128*	*304*

[a] See note to Table X.8.

electronics applications have been introduced more widely in the lower paying nationalised industries. But, whatever the reason, the pattern stands in striking contrast with private manufacturing. There were not sufficient numbers for viable comparisons between the pay of women in different circumstances in the nationalised industries. Microelectronic applications affecting manual workers were also too rare in public services for satisfactory comparison to be made between places affected and those not. The pattern in public services was more revealing so far as non-manual applications were concerned.

Earnings and new technology in the office

As Table X.9 shows, employees in establishments with computers or word processors tended to earn more at every level than counterparts in places with none. This was true of both men and women. The difference remained marked when we compared establishments of similar size, although there was a limit to which we could take that analysis, as there were so few larger workplaces which had no computing or word processing facility. The striking feature about computing and word processing applications compared with the use of advanced technology at the manual level, however, was that the differential appeared to apply equally to every grade rather than being especially marked for particular categories. The pattern suggests that whether or not establishments had computers or word processors was more a measure of the general resources of workplaces than a feature which had special implications for the earnings of particular categories.

The most striking feature of the pattern of earnings in relation to the distribution of computing and word processing, however, lay in differences between the broad sectors. In private manufacturing, all categories earned more where computers or word processors were used, although the differential was very much larger for women. In private services there was a substantial differential for both genders. In public services, however, the pattern was remarkable. Men who worked at places with *no* computing or word processing facilities generally earned more at every level than men working at places which had them. Among women, the pattern was more mixed, but for semi-skilled and skilled manual workers and for clerical and secretarial staff the differential was in the conventional direction.

Earnings in relation to combinations of manual and office usage

When we looked at earnings in relation to whether establishments operated advanced technology affecting both manual and office workers, or just one of the applications or neither, we found the pattern shown in Table X.10. There was some suggestion throughout our analysis of earnings in relation to the applications of new technology that the financial benefit derived from working at a place using advanced technology was greater for women than for men. This suggestion was supported by the analysis in Table X.10. There was, among women, a clear gradient in levels of earnings in relation to the extent to which places used advanced technology. In places where there were both manual and office applications, women earned most in every grade. In places where there was neither, women earned least. Cases where there was only one type of application occupied an intermediate position, although those where there were office applications but not manual ones were slightly more favoured than those where there were manual applications but no word processors or computers. We had difficulty in testing how the overall pattern for women persisted within size bands owing to the small numbers in the different cells in certain size bands. But where comparisons were possible they showed that the pattern was independent of variations in size. So far as men were concerned the overall trends were less clear-cut. As was the case among women, men working at places where there were no applications earned least in every grade. The trend for skilled manual workers and for supervisors was similar to that for women. So far as other grades were concerned, it appeared that men fared best who worked at places where there were office applications but no manual applications. Again the pattern for men persisted when size variations were taken into account.

We have reported in the second half of this chapter our analysis of gross weekly pay, first, in relation to whether or not manual workers operated processes including microelectronics, secondly, in relation to whether workplaces used computers and word processors and, thirdly, in relation to combinations of usage of new technology affecting manual and non-manual workers. We also carried out the analysis using our categorisation of different forms of change as our measure of the use of advanced technology and using *hourly rates* as our measure of pay. We found that, whatever measures we adopted, there was generally a positive association between use of the new technology and level of earnings.

Footnote

(1) There are, of course, the long-standing associations between both high productivity and capital intensity and relatively high wages. But it is instructive that three recent reviews of research on the introduction of new technology and industrial relations make little or no mention of the implications of advanced technical change for pay. See Jim Northcott *et al.*, *Chips and Jobs: Acceptance of New Technology at Work*, PSI, No.648, 1985; Paul Willman, *New Technology and Industrial Relations — a Review of the Literature*, Department of Employment Research Paper, No.56, 1986; Peter Senker (ed.), *Learning to use Microelectronics: A Review of Empirical Research on the Implications of Microelectronics for Work Organisation, Skills and Industrial Relations*, Science Policy Research Unit, University of Sussex, 1984.

XI Summary and Conclusions

'We all know how profound and pervasive is resistance to change at every level of our social structure. We have been brought up on the literature'. So said a distinguished social psychologist at a meeting I attended when drafting these conclusions. In the light of the results of our present research it appears that the literature was wrong or has become out of date, so far as *technical change* is concerned. Certainly, the stream of thinking that suggested that workers and their trade unions could be expected to be resistant to technical change has been shown to be inadequate by our results. In summary we found that British workers generally experience and accept a very high level of major change. Major change affecting manual workers is common, and change affecting office workers is even more so. We found no evidence in our study of 2,000 workplaces that the rate or form of change affecting either category of employee was inhibited by trade union organisation. Technical change was generally popular among the workers affected, but both shop stewards and full-time union officers tended to support changes even more strongly than the ordinary workers they represented. Where there was resistance to change by workers and union representatives, it was most frequently provoked by organisational changes introduced independently of any new technology. Even so, there was no sign that the general effect of trade unionism was to act as an obstacle to any form of change. On the other hand, it did appear that systems of management or the nature of management structures *were* associated with the propensity of workplaces to innovate. So great has been the support of workers and trade union representatives for technical change that managements have not had to use consultation, participation or negotiation to win their consent to change. Even major changes have been introduced with surprisingly little consultation. It has been very rare for technical change to be subject to joint regulation. Both manual workers and office workers have tended to derive benefit from technical change in terms of increased earnings and more skilled, more responsible and more interesting jobs. The workers directly involved in the introduction of advanced technical change have tended to be insulated from any damaging consequences of the change as regards employment. Managements have developed methods of reducing the size of workforces that are relatively painless.

The Rate of Change

We distinguished between three forms of major change: *advanced technical change*, which involved the adoption of new machines, plant or equipment incorporating microelectronics; *conventional technical change*, defined as the introduction of new machines, plant or equipment that did not include microprocessors; and *organisational change*, which meant the restructuring of working methods or working practices without bringing in new machines or other forms of hardware. Over one half of manual workers were employed at places that introduced major change in the previous three years. One third worked at places that experienced advanced technical change over that period. But both manual workers and major change affecting them were concentrated in manufacturing industry. In manufacturing, over three quarters of manual workers were employed at places that experienced major change and over one half worked at places that had introduced advanced technology in the recent past. In many instances, workplaces experienced a number of different forms of change affecting manual workers. They introduced changes in organisation and new conventional technology as well as new advanced technology. The idea that industrial workers live with constant change was given strong quantitative support by our results. Our analysis also showed, however, that the availability of relatively low cost mini and micro computers and word processors, made possible by the development of microelectronics, has led to a rate of change in the office that has been even greater than that on the industrial shop floor. Moreover, it has affected all sectors of office employment. Historically, the principal focus of interest in the human implications of technical change has been industrial workers in manufacturing industry. Now it is apparent that office workers experience such change even more frequently. Indeed, overall, twice as many workplaces introduced advanced technical change in the office as introduced applications affecting manual workers. Across all sectors of employment, nearly three quarters of non-manual workers were employed at workplaces that recently made some major change in the office, and nearly two thirds worked at places that made changes in the office associated with the introduction of computers or word processors or both. Moreover, even though change in the office was dominated by microelectronic developments, office workers also experienced other forms of major change with a level of frequency similar to that among manual workers.

Apart from specific changes of the three types we identified, the large majority of workplaces experienced changes in the size of their workforces. Only nine per cent of all workplaces had a payroll of similar size to their complement four years previously. In some instances, workplaces grew larger; in more cases they were smaller. Forty-one per cent of places reduced the number of their employees in the previous year. Reductions in the size of workforces tended to be particularly common in places where unionisation among manual workers was high, which had clear implications for the future development of manual trade unionism.

Rates of Change and the Use of Advanced Technology in Relation to Trade Union Organisation

As a result of the recent period of rapid technical change, large proportions of both manual and office workers were employed at places that used advanced technology affecting their category of employee. This was the case for 36 per cent of manual workers, a proportion that rose to 64 per cent in manufacturing industry, and 85 per cent of office workers. We thoroughly explored sources of variation in the usage of advanced technology and also in the extent to which workplaces introduced different forms of change. We paid particular attention to the possibility that there might be associations between the technology in use and labour relations institutions and arrangements. As indicated at the start of these conclusions, there was no sign of any negative association between the level of unionisation at workshops or offices and the extent to which they introduced and operated advanced technology.

The two chief features of organisation through which we sought to test the impact of trade unionism were, first, whether or not trade unions were recognised for particular categories of employee and, secondly, the trade union density in the category. Overall we found that there was very little association in either direction between trade union organisation and the introduction of change. Where, however, there was any association, it tended to be positive. In other words, places where trade unions were recognised were more likely to introduce technical and other forms of change than places where unions were not recognised. Similarly, the higher the level of trade union membership at workplaces, the more likely there was to have been change. Often, we have to emphasise, the relationships were neutral but, where they were not, then the associations were in the directions we have indicated. For instance, so far as the use of advanced technology affecting manual workers was concerned, manufacturing plants that recognised trade unions were more likely to operate advanced processes than factories that did not. The difference was lessened when we compared plants of similar size, but it remained true that factories recognising manual unions were more likely to use advanced technology in all size bands in manufacturing industry where it was possible for us to carry out analysis. Similarly workplaces where manual unions were recognised were more likely to have introduced some form of major change in the previous three years. This was true across the economy as a whole, and within manufacturing industry in particular. It was also true for change involving advanced technology and for other forms of change. When we looked at the introduction of change and the use of advanced technology in relation to manual union density, we found no such clear pattern. In general, the results were neutral, although there was some indication that smaller manufacturing units were more likely to introduce major change the higher the level of trade union membership among manual workers.

So far as non-manual union recognition and the introduction of change into the office and the use of computers and word processors in offices were concerned, the pattern was more mixed than on the shop floor. It was certainly the case that, within all three major sectors, places that recognised non-manual unions were more likely to have computing arrangements, but this difference was largely the consequence of larger workplaces being more likely both to have computers and to recognise non-manual unions. When we continued the analysis within size bands, the pattern became mixed. The chief feature of the relationship between non-manual trade union density and the use of computers and word processors was that usage tended to be highest in places where there was an intermediate level of union membership, and lowest in places where membership was sparse or dense. In particular, there was a slight indication that places with a non-manual closed shop tended to have an especially low level of usage of advanced office technology. On the other hand, they also tended to have experienced an unusually high level of organisational change. Indeed, a feature of our analysis of non-manual trade unionism and change was that a relatively high level of non-manual unions tended to be associated with a relatively high level of organisational change. As we observed in the body of the report, all our evidence combined to show that organisational change was very much less popular than technical change and it was striking that it should be so common in places where non-manual unions were relatively strong. We did not conclude from this association that non-manual trade unionism positively encouraged organisational change any more than we concluded that the recognition of manual trade unions positively encouraged the use of advanced technology on the shop floor. There were good reasons to suppose that in some instances they did, as we discuss a little later. But first we examine some further evidence of the extent to which trade unions facilitated the introduction of advanced technology.

Worker and Trade Union Support for Advanced Technical Change

When commentators, especially managerially-orientated commentators, discuss the reactions of workers and trade unions to technical change, the phrases most frequently used tend to be *worker resistance to change* and *union resistance to change*. As we quoted in the body of the book, some have identified such resistance as one of the main causes of the relatively poor performance of the British economy compared with our international competitors. Resistance to change among manual workers and manual trade unions has most frequently been singled out for special criticism. We gained a very different impression from our discussions with general managers, personnel managers, works managers and shop stewards about the reactions of the workers affected, and of any trade union officers who were involved, to the most recent advanced technical change introduced at their workplace. From these reports it was clear that the general reaction was support for change, and often enthusiasic support. This was the case on the shop floor as well as in the office. Indeed, the picture we found suggested that commentators should replace their stock phrases with references to *worker support for change* and *trade union support for change* when talking about worker and union reactions to technical change. The picture was fully consistent with our conclusion that trade unions did not operate as an obstacle to technical change.

Overall, our results showed, first, that, when it came to the introduction of particular changes involving advanced technology at the workplace, the general reaction of the workers affected was favourable. Secondly, in the cases where either shop stewards or full-time officers became involved, they tended to support the change even more strongly than their members. Where there was any resistance from either shop stewards or full-time officers to the introduction of advanced technology, then that resistance tended to reflect even greater reservations about the change among the workers affected. Certainly our findings combined to show that, in practice, both stewards and full-time officers tended to be even more strongly in favour than ordinary workers of technical changes in the office and on the shop floor.

When our principal management respondents were asked about the reactions of the manual workers affected to the most recent change at their workplace involving advanced technology, they reported that *workers supported the change in three quarters of all cases. In 37 per cent of cases they said that the support was strong. They reported strong resistance in only two per cent of cases.* The accounts of works managers and shop stewards were consistent with those reports. So far as the introduction of computers and word processors in the office was concerned, managers reported levels of support from office workers that were even higher than those on the shop floor.

Superficially, from the accounts of managers it appeared that support for change in the office and on the shop floor was slightly less widespread among shop stewards than among ordinary workers, and even less widespread among full-time union officers. This impression was misleading, however, for two reasons. First, shop stewards were more likely to become involved in cases if there were some problems involved. Full-time officers were involved in only a small minority of all cases, and these were the ones where the reactions of the workers affected were least favourable. Indeed, when we confined analysis of cases affecting manual workers to those where shop stewards were involved, then, even according to the accounts of managers, stewards tended to support the change more than the workers affected. Similarly, when we confined analysis to those cases where full-time officers were involved, then officers more often favoured the change than stewards or workers. Indeed, the number of cases in which full-time officers were involved were so few and the level of their support for change was so great that, overall, there was more resistance to advanced technical changes from first-line managers than from full-time trade union officers. We should emphasise that this contrast was chiefly a consequence of the fact that the reactions of first-line managers were generally felt by our principal management respondents to be relevant in the very large majority of cases of advanced technical change, while the stances of full-time trade union officers were felt to be relevant in only 28 per cent of cases affecting manual workers. Nevertheless, the fact that there was resistance to change from first-line managers in more cases than there was resistance from full-time trade union officers does place very much into perspective the role of full-time trade union officers in the introduction of advanced technical change. This was especially so when it remained true for advanced technical changes on the shop floor in manufacturing workplaces that recognised trade unions. Full-time trade union officers were involved in technical change in the office even less frequently than they were on the shop floor. Indeed, their views were felt by managers to be relevant in only 12 per cent of office cases. But where they were involved, their reactions compared with those of their members were generally similar to those on the shop floor. An exception was provided by the public sector, which we consider when discussing reasons for the comparatively poor performance of that sector in adopting new technology.

Secondly, the pattern of reactions summarised so far arose from the accounts of managers. When we focussed upon the separate reports of shop stewards of their own feelings towards the change and of the feelings of the workers affected, we found that the picture which emerged consistently revealed stewards as more strongly in favour of advanced technical change than the workers. In our discussions with shop stewards we were able to distinguish between initial reactions to the proposed change and subsequent feelings about it, after, in most cases, it was implemented. We discuss some of the implications of this distinction in the next section. Generally, however, stewards reported that both they and the workers affected supported the proposed technical change from the start and that they supported it more strongly than the workers. For instance, one half of our non-manual stewards were *strongly* in favour of the most advanced technical change at their workplace from the start, when feelings were most lukewarm and when their enthusiasm was shared by only one third of the office workers affected by the change.

In view of the way in which resistance to technical change among *manual* workers and trade unions has so frequently been singled out as a major problem for British industry, it is worth highlighting the results of our comparison between worker support for change in the office and support on the shop floor. Generally, levels of support were very similar, as was the pattern of variation as between workers and trade union representatives. The chief difference was that slightly more support for office change was reported by both managers and shop stewards. But the differences were small. Moreover, as we explain more fully in a later section where we focus upon differences in both the content of technical changes affecting office workers compared with changes affecting manual workers, and in the methods of their introduction, the only surprise in the contrast between the respective reactions of the two groups lay in the fact that it was not more marked.

Qualifications to the extent of worker and trade union support for technical change

In general, the picture of worker and trade union support for technical change, including advanced technical change, was strong, and contrasted markedly with the conventional wisdom. But this picture needs to be qualified in three ways. First, while advanced technical change was generally welcomed, there were pockets of resistance. The overall indications were that they tended to be concentrated in particular types of workplace, especially large establishments and those belonging to nationalised industries. Both managers and shop stewards agreed that there was less support for technical change affecting manual workers in places employing 500 or more manual workers and, independently of size, there was less support in nationalised industries than in any other sector(1). So far as office applications were concerned, the pattern in the public sector was more complex, as we discuss when we review the generally poor performance of that sector in adopting new technology.

Secondly, our present findings confirmed the results of our previous research showing that there were marked differences in the reactions of workers and stewards to change at different stages of its introduction(2). In this study we were able to distinguish only between initial reactions to the news of the proposed change and subsequent feelings at the time of interview, in most cases subsequent to the implementation of the change. This was sufficient to demonstrate, however, that there was a marked change of view between these two points among both the workers affected and stewards according to the accounts of works managers, manual shop stewards and non-manual shop stewards. We were asking about change retrospectively, at a point in time when many would have been in the frame of mind most favourable to the change (when any doubts had been stilled; when they were beginning to feel any benefits of the change in terms of any increased earnings and job enrichment, as described in later sections; and before they became habituated to any such benefits). In consequence, retrospective accounts of initial reservations may have understated actual feelings at the time or certainly given a more favourable impression of reactions than those we would have gleaned had we talked to people at that time. Thirdly, while there was no doubt that support for conventional technical change was strong and support for advanced technical change was only slightly less so, reactions to organisational change were very much more mixed. Indeed, according to many accounts, the initial responses to news of proposed organisational changes tended to be hostile, on balance. This takes us on to one of the major sets of findings in our present study: the contrasts in the impact and implications of the three different forms of change that we studied.

The Popularity of Technical Change Compared with the Unpopularity of Organisational Change

As we explained in Chapter IV, previous research on the management of change led us to expect that the chief influences upon reactions to the change would be, first, the content of the change and, secondly, the method by which it was introduced. By the content of the change we meant its implications for earnings, for employment, for intrinsic job satisfaction and for physical working conditions. By the method of its introduction we meant principally the extent to which, and the methods through which, the workers affected by the change and their representatives were involved in the change. We did not expect that the form of the change would have an independent influence upon its impact and implications. We distinguished between, and asked about, *organisational change* and *conventional technical change* as well as *advanced technical change* in the expectation that we would find few intrinsic differences among them. We expected to find that the introduction of advanced technology was only one form of change among many, and that it was not very different from other familiar forms of change. Insofar as there were different reactions to that form of change, we expected that they would result from any different implications of the change for earnings, employment, job content or working conditions, or from any different methods through which managers sought to introduce this form of change because they conceived of it as being different.

We certainly found that there were only minor differences in reactions to advanced technical change compared with conventional technical change. But there were major differences in the widespread support for both types of technical change compared with the relative unpopularity of organisational change. The reports of general managers, personnel managers, works managers and shop stewards combined to show that both workers and union representatives tended to be in favour of technical change at every stage. In contrast, the balance of feeling about organisational change was generally hostile, especially when it was initially proposed. For instance, *works managers reported that manual workers were in favour from the start of three quarters of advanced technical changes in manufacturing industry. In contrast, they also reported that in three quarters of the organisational changes in that sector, workers were initially opposed to the proposals.*

It is also true that we found, as we expected, that where changes led to loss of earnings, loss of jobs or less interesting work there was more resistance. In this context, however, there were indications that workers attached more importance to earnings and intrinsic job satisfaction and less to the implications for manning, while trade union officers attached a higher priority to the implications of the change for employment. So far as the method of introduction was concerned, consultation and negotiation tended to be responses to initial resistance to change rather than channels through which change was better designed and made more readily acceptable, as we explain more fully in a later section. Hence, for example, there was much more consultation and negotiation over unpopular organisational changes than over popular technical change.

Our definition of organisational change was *substantial changes in work organisation or working practices not involving new plant, machinery or equipment*. The particular forms of change reported under this heading were often productivity agreements of the classical type and, more generally, they appeared to be the type of change that was introduced under the productivity criterion of the incomes policies of the late 1960s and early 1970s. Often they involved reductions in manning, and they contributed to job enrichment less frequently than technical change. Moreover, organisational changes affecting manual workers were more often associated with loss of earnings. It was clear that these characteristics contributed to the relative unpopularity of organisational change. Even when we confined our analysis to cases which had similar implications for jobs and earnings, however, we found that organisational change remained very much less popular than technical change. *Indeed, technical change associated with loss of jobs was more popular than organisational change in a climate of growth in jobs. Not only was the form of change important independently of its implications for employment; it was more important than the implications of the change for employment.* The profound significance of this finding took on even greater interest when we added our analysis of the way in which the introduction of advanced technology in manufacturing industry was associated with more flexible systems of working, especially with the development of multi-skilled craftsmen and enhanced craftsmen, as we discuss later in these conclusions.

We have seen that part, but by no means all, of the reason for the relative unpopularity of organisational change compared with technical change was its less favourable content. We are not able to say, however, on the basis of the information that we collected in our present research, why organisational change should have been so much less popular independently of its content, nor indeed can we say why conventional technical change should have been consistently, if slightly, more popular than advanced technical change. But combining the pattern that we found with the results of earlier case studies of change suggests that among the attractions of new technology, independent of any more favourable content for jobs that it may have, is the fact that it represents investment, optimism, progress and achievement. Changes arising out of technological innovation tend to be seen as the requirements of desirable innovations rather than the impositions of an arbitrary authority. Where the new technology is unfamiliar it may raise slightly more apprehensions, but generally the inherent attraction of new machines is inclined to dominate reactions in our society. The rationale for organisational changes, in contrast, is often very much less self-evident. They may be seen as being imposed at the whim of neutral or unsympathetic managements. Particularly striking in our results was the way in which organisational changes were frequently initiated at levels within organisations higher than the places at which the people affected worked, and the way that workers at places belonging to large organisations with many operating units were substantially more subject to organisational change.

Whatever the reasons for the inherent unpopularity of organisational change compared with technical change, however, the distinction between the three forms of change served as a major source of variation throughout our analysis. Indeed, it was often the major source of variation. When we looked at issues such as the extent of negotiation over change, the extent to which shop stewards consulted full-time officers over change, the extent to which professional personnel managers were involved in the management of change and the stages at which they were involved, then we found that there tended to be larger differences between the practices adopted in relation to the three different forms of change than between the practices adopted by different types of workplace in relation to the same form of change. In other words, it mattered more whether workplaces were introducing advanced technical change or conventional technical change or organisational change, than whether the workplace was a large or a small one, whether it was in the public or private sector, whether it recognised trade unions or not and whether it was overseas or domestically owned. Clearly, our expectation was not fulfilled that advanced technical change would prove to be just one form of changes among many and that it did not differ intrinsically from others. But the distinctions we made nevertheless produced novel and important results. Organisational change emerged as the most distinctive form and the one that generated most problems and resistance. In contrast, both forms of technical change frequently appeared to be little more than routine. In cases where they were not, workers tended to derive immediate benefits and to be cushioned from short-term costs, as we go on to show when we summarise the implications of technical change for earnings, for the intrinsic rewards of work and for jobs. First, however, we look at the pattern of variation in the use of advanced technology in relation to sector, type of employer and, by implication, management structure.

Differences in the Use of Advanced Technology in Relation to Type of Employer and Management Structure

While we found no evidence of any general tendency for the use of advanced technology or the introduction of change to be inhibited by trade union organisation, we did, as we mentioned earlier, find major contrasts in the extent to which establishments used advanced technology both on the shop floor and in the office in relation to differences in type of employer. *Public services and nationalised industries were very substantially less likely to use advanced technology than private services and private manufacturing industry.* It was hardly a surprise that, for instance, manual workers in manufacturing were more likely than public service manual workers to operate machines using microelectronics. But the contrast between the sectors was equally marked in relation to the use of new technology in the office. Public services and the nationalised industries were much less likely than both parts of the private sector to have adopted word processors and computers to modernise office working. *Moreover, in the private sector, there were major differences between British-owned and overseas-owned establishments in the extent to which they used the most up-to-date methods, and also between independent workplaces and those that were parts of groups or larger organisations.* The differences between the public and the private sectors, and between workplaces under different forms of ownership in the private sector, were so gross that it is worth illustrating them at this stage.

Forty-three per cent of private manufacturing establishments had manual workers operating advanced technology compared with 30 per cent of nationalised workplaces, 10 per cent of private service locations and 7 per cent of establishments in the public services. Within private manufacturing industry, nearly three quarters of factories belonging to overseas companies had a process or product application that employed microelectronic technology, while this was true for only 42 per cent of domestic counterparts. Furthermore, it was not simply that plants belonging to overseas owners were more likely to use advanced technology, they were also very substantially more likely to introduce organisational change and conventional technical change. Similarly, factories that belonged to larger companies were very much more likely than independent establishments to use advanced technology.

As implied earlier, when it came to comparisons across broad sectors, differences in the use of computers and word processors in the office were perhaps better guides to the modernity of managements than micro-electronic applications on the shop floor. This is because office operations were common to all sectors, while the activities of manual workers varied very considerably. Also, office applications of new technology generally require a lower level of capital investment than applications on the shop floor. Hence, they may depend more upon managerial innovativeness independently of capital resources. Private manufacturing industry appeared consistently more up-to-date in its office technology than nationalised industries, and private services were substantially in advance of public services. For instance, nearly one half of private manufacturing plants recently introduced change in the office involving computers or word processors, compared with 30 per cent of nationalised industry workplaces. Thirty eight per cent of private service establishments made similar changes compared with 23 per cent of offices engaged in public services.

In the private sector, the contrast between workplaces belonging to overseas companies and domestic counterparts in their use of advanced technology in the office appeared, if anything, to be greater than the differences in their usage on the shop floor. Two thirds of workplaces belonging to overseas companies recently introduced change involving computers or word processors in the office compared with 39 per cent of British counterparts. Again, overseas-owned places were also very substantially more likely to make other forms of change. The marked contrasts between the extent to which workplaces with different types of employer used advanced technology appeared to imply that the innovativeness of workplaces owed much to their systems of management or management structures. Such an inference could most legitimately be drawn from the comparison between workplaces belonging to overseas companies operating in Britain and those under British ownership. When making this contrast we were able to compare workplaces of similar size, operating in similar broad sectors and all belonging to larger organisations. Whatever the basis of the comparisons, overseas companies were consistently much more likely than British counterparts to use advanced technology both on the shop floor and in the office. They were also much more likely to introduce organisational changes in addition to, and separately from, their relatively high level of technological innovation. It became increasingly difficult not to attribute a large part of the contrasts to differences in management as between overseas-owned companies operating in Britain and domestic counterparts.

Moreover, our conclusions that variations in the rate of adoption of new technology owed much to management and little, if anything, to levels of trade union organisation are consistent with previous research. Using a quite different method, Northcott came to a similar conclusion(3). He asked managers about the obstacles to introducing advanced technology on the shop floor in manufacturing industry. Of course, there is a danger in using managers' accounts of obstacles alone that those accounts will be contaminated by managers' attitudes and they will, consciously or unconsciously, seek to explain away management deficiencies by blaming unions or workers. Even so, Northcott has found in a number of studies during the 1980s that managers have rarely cited worker or union resistance as a reason for not introducing advanced technology. The most common obstacles to change that have consistently emerged from managers' accounts have been lack of qualified or skilled manpower, lack of management confidence owing to the economic climate, lack of finance to fund investment and lack of management awareness of new developments. When Northcott's results are added to our own they strongly reinforce the weight that may be attached to the common conclusion of both studies.

Some Observations on the Variable Performance of Managements in Adopting New Technology

Our conclusion on the relative importance of management compared with the influence of trade union organisation in determining the rate at which advanced technology has been introduced in British workplaces is supported by the two main independent pieces of research that have been carried out on the subject. These were conducted separately using different methods. Our present research systematically and thoroughly plotted both the distribution of advanced technology and the rate at which different forms of major change were introduced in workplaces, in relation to trade union organisation and other characteristics of workplaces. Northcott asked managers about the obstacles to change. We both came to a similar conclusion. On the other hand, this conclusion may seem implausible in the face of what may appear to be common observation. For instance, first, there have been highly publicised examples of trade union resistance to advanced technical change in the Post Office, in British Rail and in Fleet Street. Secondly, businesses like Marks and Spencer and IBM are well known as leaders in the use of advanced methods and techniques in their fields and also well known for consciously organising their businesses to avoid the constraints of trade union representation and recognition. Our present findings strongly suggest, however, that trade union resistance to technical change and a high rate of innovation and modernity in places that do not recognise trade unions are both very much exceptions to the general rule.

The implications of our present findings are that the isolated cases where trade unions do resist technical change command public attention, but *in other places the existence of trade unions must act less publicly as a positive encouragement to the introduction of advanced technology*. This may be because union representatives actively seek new and better tools for their members; or because trade union structures represent one of the major networks through which information about new technology and its implications is disseminated (we have seen how frequently full-time manual union representatives acted as consultants to managers when they were introducing new technology); or because the existence of trade union representation and organisation acts as a stimulus to managements to substitute capital for labour, either as a result of the trade union mark-up on wages or through some other medium(4). In these ways it can readily be seen how trade unions may encourage managements to introduce change in some cases, in a manner that is very much less publicly visible than in the isolated cases where change is resisted through trade union channels.

Similarly, our findings also suggest that instances of managements which are active, innovative and progressive, and also adopt a conscious management strategy to avoid recognising trade unions for any grade of employee, are also exceptional. Generally, the impression created by our findings, overall, was that managements that did not recognise manual unions tended to be relatively passive, conservative and even apathetic. For instance, they were very much less likely to engage in consultation over the introduction of change, even informal consultations with individual manual workers or groups of workers, than counterparts which recognised manual unions. Secondly, our analysis of working practices in manufacturing industry revealed that, although managements which did not recognise manual trade unions reported many fewer constraints on their freedom to organise work as they liked, their accounts of actual working practices in their plants showed that there was less flexibility of working than in workshops where unions were recognised. Multi-skilled and enhanced craftsmen were less common; production workers less frequently carried out simple maintenance tasks; and they were very much less likely to have taken initiatives to promote flexibility in recent years.

Two further sets of observations may be made about our findings on the way in which managements of different kinds varied in their adoption of new technology. The first relates to the contrast between British and overseas-owned companies, and the second to the contrast between the public and the private sectors. First, the greater tendency for manufacturing workplaces owned by overseas companies to be much more likely than British counterparts both to employ advanced technological processes and to make products incorporating microelectronics might have owed as much to the sectors of manufacturing in which they were engaged as to their more innovative managements. Whichever was the stronger influence, however, the contrast reflects ill upon British companies. Whether their relatively poor performance in the adoption of advanced technology is a consequence of their languishing in older industries or of their failure to be as innovative as others in their own industry, the contrast is not encouraging for the future of British industry. Moreover, as we have already argued, the fact that workplaces owned by British companies also trailed behind foreign-owned counterparts in the introduction of other forms of change affecting manual workers, and in the use of word processors and computers in the office and in other forms of change affecting office workers, does very strongly suggest that the contrasts can not wholly be explained by differences in the nature of the activities being carried out at the establishment.

Secondly, however far British managements might be behind foreign-owned counterparts in the private sector, our results show that they were well ahead of British managements engaged in the public services and in the nationalised industries. They suggest that if measures are needed to improve the performance of British management, the need is most urgent in the public sector. Of course, this does not necessarily mean that individual managers in the public sector are generally inferior in personal characteristics or even in particular management skills to individual managers in the private sector. It may be that the management structures in which they operate provide them with less scope to express their abilities and use their skills. In this context it is worthwhile to summarise some of the distinguishing characteristics of the public sector in general and of the nationalised industries, in particular, so far as our findings on the introduction of change are concerned. We do not suggest that these characteristics explain the relatively poor performance of the nationalised industries and the public services in adopting new technology. In particular, they ignore the different systems of capitalisation that characterise the different sectors; on the other hand, they do represent distinctive characteristics of the nationalised industries in particular, which it would be worth looking at in seeking an explanation.

Principally, in terms of the characteristics and behaviour which we examined, the nationalised industries were distinguished by the level of centralisation in management decision-making over the introduction of change. Not only were decisions to introduce change very much more likely to be centralised, but attempts were also more likely to be made to control the implementation of change from the centre. Associated with the degree of centralisation were the relatively very extensive involvement of full-time union officers in the introduction of the change and the lack of consultation at local levels over the decision to introduce the change as opposed to decisions about its implementation. Managers reported that full-time union officers were consulted in 57 per cent of cases involving the introduction of computers or word processors in the office; the comparable figure in both parts of the private sector was four per cent. While consultation over the introduction of change at some stage of the process was relatively common in nationalised industries, consultations with people at plant level about whether to introduce the change were less common than in any other sector. The tendency for manual workers to be much less strongly in favour of technical change in nationalised industries, and for there to be a comparatively high degree of dissonance between the accounts by local managers and shop stewards of reactions to change in the office, may also have been associated with the centralisation of decision-making. The accounts by both managers and stewards of reactions to change on the part of the workers affected and of shop stewards revealed a greater contrast between the responses of stewards and those of workers than in any other sector. Moreover, while, according to the reports of managers, stewards were lukewarm about the change compared with workers, the accounts of stewards suggested that they were substantially more enthusiastic than their members. There were clear indications of very much poorer communications than in any other sector both between managers and stewards and between stewards and their members.

The second striking general feature of the use of advanced technology in the nationalised industries was the way in which the earnings of different categories did not move in line with its use. In other words, in the private sector we found that in places where advanced technology was used on the shop floor or in the office, the earnings of different categories tended to be higher than those of counterparts elsewhere. There was no evidence of any such pattern in the nationalised industries.

Advanced Technology, Job Content and Work Organisation

Our results provided support for those who have argued that the spread of advanced technology would enrich the jobs of the workers affected, certainly in comparison with the quality of work under the characteristic systems of production or working for manual workers during the first half of the twentieth century. We found little comfort for those on the other side of the debate, who have taken the view that the development of automation represents a further stage in the dehumanisation of work. *The generally favourable picture that we found of the impact of new technology upon the content of jobs applied both to manual and to office workers.*

We did not have reports from the workers themselves of their feelings about working on the new technology or about their sources of satisfaction or deprivation. But the separate and independent accounts that we had from general managers, personnel managers, works managers and shop stewards were sufficiently consistent to be persuasive. They showed, first, that where the content of jobs was changed as a result of the introduction of advanced technology, the balance of change was towards job enrichment. Levels of skill and responsibility and the range of tasks carried out by workers were increased. Intrinsic job interest was enhanced. Secondly, the changes appeared, on the other hand, to have a more neutral or even slightly negative effect upon the autonomy of workers as measured by their control over how they did their jobs, the pace of their work and the degree of supervision to which they were subject. We emphasised in our analysis that ideas about control are complex, but there was little doubt that workers derived much less benefit in terms of control than in relation to other sources of satisfaction. This appeared to be especially true for manual workers in larger workplaces. Thirdly, office workers tended to derive slightly more job enrichment from technical change than manual workers. Managers more frequently reported that the content of office workers' jobs changed as a result of the technical innovation, and in cases where there was change, it was more frequently described as favourable.

In larger manufacturing plants we were able to pursue our interest in the implications of advanced technology for the organisation of work more fully through our interviews with works managers. Our principal focus was on the distribution of tasks between different categories of worker and the constraints in the organisation of work to which managers were subject from collective agreements, the informal activities of workers and shop stewards, the technology operated in the plant, or the lack of appropriate skills among workers. We were especially interested in how far the organisation of work was changing and how this was related to the use of new technology. Our results appeared to have substantial significance for the management of change and the promotion of flexibility of working in manufacturing industry.

Our analysis provided support for the view that there has been a tendency for traditional demarcations as between different categories of manual worker in manufacturing industry to break down over recent years. For instance, production workers were increasingly becoming involved in straightforward maintenance work and the setting up of machines. Craftsmen were becoming more multi-skilled. Such changes were coming about both as a result of initiatives on the part of managers to achieve changes in working practices, and more spontaneously, as a spin-off from other forms of change. In particular, it was clear that many of the organisational changes reported in larger manufacturing plants were productivity agreements of the classical type. They frequently included provisions to promote flexibility of working as between different categories of worker. In view of our historical interest in the practice, we were delighted to discover that productivity bargaining was alive and well despite rumours of its demise a long time ago. In addition, however, we found that the introduction of advanced technology was also strongly associated with more flexible systems of working, especially the creation of the new categories of enhanced craftsman and multi-skilled craftsman. The chief difference lay in the reactions of workers and shop stewards to changes in working practice taking different forms. In particular, it was clear that they were very much more popular when associated with technical change than they were when introduced independently of any new plant or equipment. Here, it is no overstatement to describe the contrasts as dramatic. We quoted earlier, when discussing the relative popularity of different forms of change, the figures derived from works managers' accounts showing how massively workers welcomed the prospect of advanced technical change from the start and how equally massively hostile they were to the prospect of organisational change. It is difficult to imagine more marked differences in reactions, and their implications for the reform of working practices are clear.

The Impact of Advanced Technical Change on Pay

We suggested in Chapter X that a reason why the implications of advanced technical change for rates of pay and earnings have received so little attention from commentators may be because of some reluctance on the part of supporters of change to promote the notion that it carries a price tag, accompanied by reluctance on the part of sceptics about change to encourage the impression that it delivers a pay-off. Certainly our findings suggested that the introduction of advanced technology did result in higher earning both on the shop floor and in the office. This conclusion arose both from answers to our question about the implications for pay of recent specific changes and from our analysis of levels of earnings in relation to the pattern of usage of advanced technology at workplaces.

In their separate accounts of the impact of recent changes upon the earnings of the manual workers involved, there was more inconsistency between managers and manual shop stewards than in relation to any other question. We attributed this to the complexity of systems of pay for manual workers combined with a tendency on the part of the respective parties to focus upon any feature of the change that suited their case. Despite the differences, however, the general picture which emerged from the separate accounts was broadly similar. The introduction of advanced technology had no direct, short-term effect upon the earnings of the manual workers affected in the majority of cases. But in a substantial minority of cases it did result in changes in earnings, and where there were changes they were generally to the advantage of manual workers. Managers reported loss of earnings in a very small minority of cases (three per cent), while stewards did so in slightly more (nine per cent). Increases in pay were reported in about one fifth of cases and they came chiefly through regrading, enhanced earnings under some system of payment by results, or negotiated increases in rates associated with a formal collective agreement relating to the change. The chief differences that emerged from parallel questioning about advanced technical change in the office were, first, that the separate accounts of managers and non-manual shop stewards were much more alike. Secondly, both parties hardly ever reported reductions in earnings arising from office change. Each reported increases in slightly more than one case in every five. Increases for office workers came predominantly through upgrading.

Perhaps stronger evidence of the financial benefit workers derived from the introduction of advanced technology arose from our broader analysis of the comparative earnings of different groups in relation to its use at workplaces. All three main grades of manual worker earned more in places where workers operated advanced technology than in places where they did not. The average differential for the skilled grade was nine per cent among men and 14 per cent among women. All grades tended to earn more in places which used computers and word processors than in places which did not. Male office staff enjoyed an average differential of six per cent, while the average differential for women office workers in places using computers or word processors was nine per cent. When we adopted combinations of different patterns of usage, then it appeared that the more advanced technology a workplace used the higher were earnings generally. The differentials described tended to persist when we carried out analysis within different sectors and size bands and when we looked at the earnings of men and of women and of the different grades separately. The more detailed analysis revealed two notable features, however. First, it appeared that in *some parts of the public sector earnings less clearly reflected the use of advanced technology*. This was especially marked so far as the earnings of manual workers in relation to their use of advanced technology in nationalised industries were concerned, and so far as the earnings of men at every level in public services in relation to the use of computers and word processors were concerned. We noted this feature when we discussed the relatively poor performance of the public sector in adopting new technology. Secondly, *there were repeated indications that women tended to benefit more than men from the adoption of advanced technology*. In other words, the differentials among women at different levels in relation to the usage of the new technology at their workplaces tended to be more marked than the corresponding differentials among men.

Impact of Advanced Technology Upon Employment

As we emphasised in Chapter IX, the information we collected in our present study enabled us to answer only two of the main questions posed about the implications for employment of the introduction of advanced technology. First, we were able to see how far advanced technical change led directly to changes in staffing in the sections where it was introduced. This did not get us very far in answering the main question, however, as there may be time lags in the impact of new technology upon jobs and its main impact may be felt outside the sections where it is directly applied. In addition, however,we collected information that enabled us to determine the changes that had taken place in the number of people employed at most of the 2,000 or so workplaces in our survey over the previous four years. This took us substantially further than much previous evidence about the implications of new technology for jobs. We were able to explore the extent to which longer-term movements in the number of people employed in workplaces as a whole were associated with the extent to which they used advanced technology. This still left questions unanswered, as we were not able to determine how far the changes at the workplaces we studied had direct or indirect employment effects outside those workplaces, or outside the enterprises of which they were part. Nevertheless, with all their limitations, our findings provided a much stronger empirical basis than has previously been available for assessing the direct effects of new technology upon jobs in the workplaces where it is introduced and for reviewing its implications for employment at those workplaces over time.

We found, first, that in a substantial majority of cases the introduction of advanced technical procedures had no short-term impact upon the number of the people employed in the sections directly affected. Where, however, it did make an impact, jobs tended, on balance, to be lost. This was true for advanced technical change both in the office and on the shop floor. So far as applications affecting manual workers were concerned, it appeared that the introduction of advanced technology led to increases in manning in about one case in every ten but to decreases in about one case in every five. These findings are consistent with previous research about direct, short-term effects(5). In instances where manning was reduced, however, it appeared that the reduction was often very substantial. Our interviews with works managers suggested that in manufacturing industry the average decrease was in the order of very nearly one half of the affected section or sections. We did not have any measure of the scale of any reductions in staffing arising from change involving the introduction of computers or word processors into offices. But our findings on the extent to which it led to any reduction in the numbers employed, and on the nature of any such reductions, were broadly similar to our results on the short-term impact of changes affecting manual workers.

Secondly, we found that, although there were important variations between sectors, there was a very slight tendency overall for those workplaces that had introduced and were using *advanced technology affecting manual workers* to be more likely than other workplaces to be employing fewer people than they were four years previously. In contrast, the use and introduction of word processors and computers into offices tended, if anything, to be associated with growth in the size of workforces. This was certainly the case where advanced technology was used in the office but there was no application to manual workers. Unfortunately, we were not able to distinguish between the numbers employed at different levels in our analysis of changes in the size of the total workforces. In view of the general pattern of our findings, however, it would have been very surprising if we had not found that manual workers had been principally affected by loss of job opportunities arising from the introduction of advanced technology.

Although the association between the introduction of advanced technology on the shop floor and relative decline in employment was slight, its importance may have been greater than its scale, for two reasons. Places that introduced advanced technology affecting manual workers were places in which investment was being made at that level. In principle, it would be to such workplaces that the economy would be looking for growth in jobs to compensate for job loss in declining sectors where there was no investment. When, in fact, *the places that were enjoying the benefits of investment at the manual level were suffering slightly more loss of employment, then the pattern was more disturbing*(6).

Moreover, *the association we found between the use of advanced technology and a declining payroll was strongest in the private service sector*. Employment in British manufacturing industry has been subject to steady secular decline and, more recently, very sharp cyclical decline. Public expenditure has been subject to tight control and there appear to be few prospects of a large-scale increase in public expenditure to create jobs under any prospective government. When to this picture is added a tendency for the spread of advanced technology at the manual level to the private service sector to be associated with a relatively greater decline in employment, then it becomes even more difficult to see where the new jobs are going to come from. On the other hand, our findings on movements in the number of people employed in relation to the introduction and use of advanced technology at the manual level did refer to a period when the general economic climate was not buoyant. We cannot say what might happen to the associations we found over a period when there was sustained growth in demand.

Thirdly, the workers employed within workplaces using advanced technology tended to be insulated from any impact that it had upon employment by the manpower policies and practices adopted by employers to control the size of workforces. This was especially the case with regard to any immediate and short-term reduction in the need for labour arising from the introduction of advanced technology. Our interviews with works managers revealed that it was very rare for workers to be dismissed as a consequence of the introduction of such change. Generally, reductions in manning were managed through the redeployment in other parts of the establishment of people displaced by new machines. There were indications, however, that subsequently, in the longer term, the introduction of advanced technology did result in some establishments making reductions in their workforces after a time. Nevertheless, it remained true that in making such reductions it was rare for employers to dismiss people. They generally adjusted the size of their workforces through natural wastage, voluntary early retirement and voluntary redundancy schemes, and had to resort to enforced redundancy in only a small minority of instances. In consequence, it appeared, first, that there was often a time lag in any reductions in the size of workforces resulting from the introduction of advanced technology. Secondly, the people who suffered from these reductions were people seeking to enter employment rather than already employed in the workplaces introducing new technology.

Here we should emphasise, however, that the introduction of new technology played a very small part in the overall pattern of manpower reductions. Generally, where workplaces were forced to reduce their complements, it was as a result of loss of demand for goods or services, or because of cash limits or cost reduction exercises or reorganisations. Managers gave lack of demand as a reason for their most recent manpower reduction in one third of cases. They mentioned cash limits in a further 30 per cent of cases. They cited the introduction of new technology as a contributory factor in only 10 per cent of cases. The implication of the overall pattern of our findings, however, was that *the critical question regarding the implications of advanced technology for employment was how far it enabled employers to avoid taking people on in circumstances where they would otherwise have done so, rather than how far it led directly to reductions in the workforce.*

Fourthly, there was no doubt from our findings that the introduction and use of advanced technology had more implications for the nature than for the number of people employed. In other words, we found that places that were using advanced technology at the manual level were substantially more likely to have been recruiting more *skilled* manual workers in the recent period. The implications of introducing computers or word processors in the office appeared to have an even more substantial impact upon patterns of recruitment. Places that used computers or word processors were twice as likely as others to have taken on secretarial, clerical or administrative staff. They were also twice as likely to have taken on supervisory, technical, professional and management grades. The clear implications of advanced technology for the composition of workforces again underline the importance of appropriate methods of manpower redeployment, if any unfavourable consequences of new technology for employment are to be reduced.

Levels of Consultation and Negotiation over the Introduction of Change

Our analysis produced two major findings so far as consultation and negotiation over the introduction of advanced technical change were concerned, particularly change affecting manual workers. First, the levels of consultation and negotiation were remarkably low. Secondly, consultations with manual workers or their representatives tended to take place only when there were representative institutions that required consultation or when there was initial resistance to the change or when some outcome of the change was manifestly unfavourable.

In view of the extent to which all parties and all representative groups in industry, commerce and administration have adopted the view that worker participation in change is desirable, it was remarkable that, *even according to the accounts of managers, shop stewards were consulted about any stage of the introduction of advanced technical change affecting manual workers in only 39 per cent of cases.* This proportion rose to only just over one half when analysis was confined to places where trade unions were recognised. Discussions in established joint consultative committees or specially constituted *ad hoc* joint committees were held even less frequently. There were informal discussions with individual workers directly affected by the change in 58 per cent of cases and with groups of workers in 38 per cent of cases. No discussions were held with any manual worker or any worker representative at any stage of the introduction of advanced technical change in nearly one fifth of cases. Moreover, as we stressed initially, this picture emerged from reports by managers of any discussions that took place with workers or their representatives at any stage of the introduction of the change. When we added to the pattern the reports of shop stewards, there emerged an even less favourable picture regarding the extent of consultation. In addition, we were able in our interviews with stewards to distinguish between consultations about whether the change should be made and consultations over its implementation. *Discussions were held with manual shop stewards both before the introduction of the change and during its implementation in only one fifth of cases where manual unions were recognised*, according to our interviews with stewards. It was in relation to issues concerning whether the change should be made that consultations were especially infrequent. This was particularly true in the nationalised industries, the sector in which formal arrangements for joint consultation with representatives of manual workers were, in principle, most highly developed. In practice, it appeared that opportunities for a local input at an early stage into decisions about change were comparatively rare. In general terms, the private service sector provided least scope for manual workers or their representatives to be involved in decisions about the introduction of major change affecting them. Even in private sector establishments employing 200 or more manual workers, managers reported no consultations with anyone at any stage of the introduction of advanced technical change in very nearly one half of cases.

In many ways, our findings on the extent of consultation over technical change showed that it was much less common than has been suggested by the limited evidence that existed previously(7). The explanation for this contrast is that the previous evidence was partial in one of two different ways. Either it was based upon case studies carried out in relatively highly organised workplaces or it was derived from survey data in one sector, based upon one simple question asked of managers only. If the extent of *consultations* over the introduction of a new form of technology which generally affected a substantial proportion of manual workers was surprisingly low, the infrequency of *negotiation* over the introduction of change was even more remarkable. This was certainly the case when the pattern we identified was placed in the context of the public debate during the time when people first started to become aware of the possible implications of the new technology. At that stage, many took the view that it was not going to be feasible to introduce it without establishing new technology agreements that would lay down procedures to be adopted at each stage of the process in order to regulate the change jointly. In practice, we found that, taking into account the reports of managers and shop stewards, only about one in every ten cases of advanced technical change affecting manual workers was subject to joint regulation in places where trade unions were recognised. Negotiations over advanced technology was slightly more common than over conventional technology, as we have already emphasised, but very much less common than bargaining over forms of organisational change that have been around for a long time. Even in large manufacturing plants with 100 per cent trade union membership among manual workers the introduction of advanced technical change was jointly regulated in only about one quarter of all cases affecting manual workers.

Perhaps even more important for public policy than the generally low level of worker participation in the introduction of technical change affecting manual workers revealed by our findings were the circumstances under which participation was more common and those under which it was less common. Principally, we found that where manual unions were recognised, then consultation was very much more common. *Where manual unions were not recognised, managers reported no consultations over the introduction of advanced technical change, not even discussions with individual workers affected, in nearly one half of all cases.* The corresponding proportion in places where manual unions were recognised was eight per cent. Moreover, it was not simply that formal joint consultations were more frequent when unions were recognised. Informal discussions with individual manual workers and with groups of workers affected by the change were also very much more common. This pattern was especially striking since dealing with workers as individuals or dealing with them collectively within the framework of trade union representation have often been presented by managers as alternatives. In the context of the introduction of technical change, it appeared that the general effect of trade union representation was to encourage managers to talk with workers on an individual basis about the prospect of changes that affected them. Similarly, joint regulation through trade union representation and joint consultation through elected works councils or similar bodies have also often been seen as alternatives. Again, we found that joint consultation through elected joint bodies was more rather than less frequent in circumstances where manual unions were recognised. The implication of the pattern relating to manual workers that we found overall was that the framework of trade union representation served as a context which encouraged managers to engage more in all forms of consultation, informal and formal. Possibly, this may have been because, even though the change was not formally subject to joint regulation in the large majority of cases, managers were aware that unless compliance with change was achieved through informal means it might move into the area of collective bargaining. Clearly, our findings on the way in which trade union representation contributed to more worker involvement in change of all types have profound policy implications for those who wish to promote worker participation.

The tendency for managers not to make arrangements for the participation of manual workers in change unless pressure of one kind or another was exerted upon them was further illustrated by the way in which consultation of all types tended also to be more common in cases when there was initial opposition to change from among manual workers or their representatives, or where some feature of the change was unfavourable to management. It may be objected that this pattern was hardly surprising, and that there will always inevitably and necessarily be more consultation over less popular changes which generate opposition. The difficulty with this objection, however, is that we did not find the same pattern in relation to consultation over change in the office. Managers seemed substantially more likely spontaneously to involve office workers in the implementation of change, independently of the popularity of the change or of formal arrangements for the representation of office workers. We discuss this feature of our findings more fully in our section on contrasts between change in the office and change on the shop floor. In this context, its importance lies in the way it serves to show the distinctive nature of the approach of British managers towards the involvement of manual workers in change. It has become fashionable of late to seek to identify distinctive management styles or orientations towards industrial relations. The impression created by our findings is that the pervasive management approach to the introduction of change affecting manual workers was quite simply *opportunist*. When managers wanted to make changes they simply set about introducing them, and if they could get away with it without consulting anyone they did so. Again, it might be objected both that this is exactly what one would expect and that it represents a perfectly reasonable pragmatic approach. This objection once again ignores the contrast with the approach to change in the office. It also ignores the benefits that may be derived from worker involvement in change, quite apart from the simple avoidance of overt resistance to the change. These benefits may include improvement in the quality of a whole range of decisions affecting the change through an input of the ideas and perspectives of the people who are to operate it; an enhanced commitment from them to the change and a more ready adaptation to it; and the promotion of generally more positive and co-operative industrial relations in the workplace(8); quite apart from any principled considerations concerning the civil rights of workers which, as we have emphasised, now appear to be shared by all parties and representative groups in industry, commerce and administration.

The small role of professional personnel management in the introduction of technical change

One reason for the poor level of consultation over the introduction of technical change may have been the failure of general management to involve professional personnel managers more in the process. We have noted how researchers looking at the introduction into industry and business of mainframe computers in the early 1960s were struck by how infrequently personnel managers were consulted. It appears that little has changed. We were able to plot the role of personnel managers only in relation to the introduction of change affecting manual workers in larger manufacturing plants through our interviews with works managers. Even in these circumstances, personnel managers were involved in the introduction of advanced technical change in only one half of all cases and they tended to be brought in fairly late in the day. There was a striking contrast between their roles with regard to technical change compared with the much less popular organisational changes. Personnel managers were most heavily involved and involved at the earliest stage in the introduction of organisational change. They were least frequently consulted about the introduction of conventional technical change. Indeed, the pattern of variation was very similar to that in consultations with workers and their representatives over the introduction of change. This was generally true of the pattern of personnel involvement in change.

Personnel managers tended to be brought in when there were difficulties over the acceptance of change. In the absence of overt problems, any contribution that they might have been able to make to the more effective management of change and use of human resources tended to be neglected. The failure to involve professional managers in the introduction of technical change was made even more remarkable by the fact that we found clear evidence that *change was more readily accepted when personnel managers were involved at an early stage.*

Contrasts Between Advanced Technical Change in the Office and on the Shop Floor

The most striking feature of our analysis of the introduction of changes affecting office workers compared with those affecting manual workers was the similarity between the separate pictures that we identified, certainly in relation to the aspects of change that we were able to measure in our present enquiry. In consequence, we have not laboriously spelled out in our conclusions minor differences in detail between the two pictures at each point, nor have we repeatedly emphasised, where we have not said otherwise, that the picture applied to both categories. Where marked differences occurred we have drawn attention to them, but where we have not specified the category we were talking about then it may be assumed that we were talking about both. In this section of our conclusions we highlight those differences that did emerge from our analysis of the ways in which technical change was introduced into the office and its impact and implications compared with the picture for manual workers.

Perhaps the most marked element in that contrast was the extent to which advanced technical change has been more common in the office. We quoted the figures demonstrating this difference at the beginning of our conclusions. They are of significance because, traditionally, interest in reactions to technical change and in the impact of technical change has focussed upon change affecting manual workers in manufacturing industry. Moreover, we also found some evidence that the impact upon jobs of advanced technical change in the office was more profound than its impact on the shop floor. Certainly according to the accounts of managers, changes in the office tended more frequently to lead to increases in skill, responsibility, task repertoire and job interest. We were not certain exactly how much reliance to place upon the scale of this difference as reflected in management accounts. It appeared clear that, to some extent, managerial descriptions of the effect of changes upon job content reflected underlying negotiating positions and the ways in which skill, responsibility and task repertoire were linked to grading systems. Local collective bargaining over pay was less common for office workers and, in any case, there appeared to be a slight managerial bias in favour of office workers. Of course, we had stewards' accounts of the impact of changes upon job content in only those cases where trade unions were recognised and stewards were interviewed. Consequently, there were reports from non-manual stewards in a smaller proportion of cases than from manual stewards. But the comparison of the accounts of the respective stewards did not reveal the same degree of contrast as the management reports.

Apart from contrasts in the extent of change and in its possible impact upon job content, our comparison of the implications of change for office workers revealed differences from those for manual workers that were surprisingly slight. Formal consultations with stewards and union officers over the introduction of change and negotiations over change were more common on the shop floor, but this was principally because manual workers were more frequently represented by trade unions. When analysis was confined to places where trade unions were recognised, then the pattern of formal consultation and negotiation was similar for both categories. More strikingly, informal consultations, especially consultation with individuals, were much more common in the office and we take up that theme a little later. Support for technical change was even stronger in the office than on the shop floor, though the difference was not marked. Office change rarely, if ever, led to reductions in earnings for the workers affected, while such reductions were not uncommon among manual workers. Immediate, short-term reductions in staffing for the section or sections affected followed upon office change with similar frequency to such change on the shop floor, though, more importantly, our overall analysis of patterns of manpower adjustment and recruitment was comparatively favourable to white-collar workers. Decisions to introduce change in the office tended more frequently to be localised, though we do not know how far this was simply a consequence of any tendency for them to cost less.

Four further observations may be made about the impact of change upon office workers compared with manual workers. First, the contrast between the two groups was paradoxically greater when the comparison was based upon the same workplaces. In other words, when we confined the analysis to those establishments that introduced advanced technical changes affecting both manual and office workers, separately, we found a more marked difference between the treatment or experiences of the two groups than when we compared the treatment and experiences of all office workers and all manual workers subject to technical change. As we repeatedly emphasised in our report when making these comparisons, the places where both groups experienced advanced technical change tended to be concentrated in larger manufacturing plants. It appeared that it was in such plants that the differential treatment of office and manual workers was most marked. As, however, the number of large manufacturing sites is in secular decline, as well as having been subject to sharp cyclical decline in recent years, we can expect that the differential treatment of the two groups may gradually decrease.

Secondly, *any managerial preconception that office workers are inherently more favourably disposed to technical change than manual workers was not supported by our findings*. It is true that, overall, our results tended to show that office workers did support technical change slightly more strongly than manual workers. But in view of the way in which they were more fully consulted about change on an individual basis and the way in which change more frequently led to favourable outcomes in terms of pay, job enrichment and employment, it was hardly necessary to invoke any more compliant predisposition to change in order to explain the modest difference in the strength of office worker support for change compared with that among manual workers.

Thirdly, the extent to which office workers are more frequently consulted over change on an individual basis was very striking, as was the way in which they needed less than manual workers to be represented by trade unions or to resist change in order to be heard. On the face of it, there would appear to be no reason why a semi-skilled worker in an office should have any more to contribute to discussions about proposed changes or should have any more right to have a voice in such discussions than a semi-skilled manual worker. But it appears that the British class problem persists and British managers still have difficulty in communicating with British manual workers in a way that, for instance, their Japanese counterparts in Britain do not experience(9).

Fourthly, and perhaps most importantly, our analysis of longer-term changes in the number of people employed in relation to the extent to which workplaces used new technology revealed clearly that the changes occurring were operating to the advantage of office workers. The implication of this longer-term trend for the comparative job security of manual workers may be exacerbated by differential terms and conditions of employment applying to the two broad grades. We did not explore this type of difference in our present study but information on the composition of people coming into unemployment suggests that manual workers have been harder hit by the whole range of changes that have been taking place over recent years(10).

Footnotes

(1) These pockets of resistance were similar to those identified in earlier reviews of the field. See, for example, Jim Northcott *et al.*, *Chips and Jobs: Acceptance of New Technology at Work*, PSI, No.648, 1985.

(2) W.W. Daniel, *The Right to Manage*, Macdonald, London, 1973.

(3) Jim Northcott with Petra Rogers, *Microelectronics in Industry: What's Happening in Britain*, PSI, No.603, 1982; Jim Northcott and Petra Rogers, *Microelectronics in British Industry: The Pattern of Change*, PSI, No.625, 1984; Jim Northcott, *Microelectronics in Industry: Promise and Performance*, PSI, No.657, 1986.

(4) David Blanchflower, 'Union relative wage effects: a cross-section analysis using establishment data', *British Journal of Industrial Relations*, Vol. 22 No. 3, 1984.
(5) Northcott with Rogers (1982), *op.cit.*; Northcott and Rogers, (1984), *op.cit.*
(6) See also Roy Rothwell and Walter Zegweld, *Technical Change and Employment*, Frances Pinter, London, 1979.
(7) Paul Willman, *New Technology and Industrial Relations — a Review of the Literature*, Department of Employment Research Paper No.56, 1986; Peter Senker (ed.), *Learning to use Microelectronics: A Review of Empirical Research on the Implications of Microelectronics for Work Organisation, Skills and Industrial Relations*, Science Policy Research Unit, University of Sussex, 1984.
(8) Senker (1984), *op.cit.*
(9) Malcolm Trevor and Michael White, *Under Japanese Management: The Experience of British Workers*, Gower, Aldershot, 1983.
(10) W.W. Daniel, *The Unemployed Flow*, Stage I Interim Report, PSI, 1981.

Appendix A

Questioning on Technical Change from the Interview Schedule

We reproduce here the questions from the interviews with principal management respondents about the use and introduction of new technology. The first section focuses upon manual workers and the second upon office workers. Similar sections were also included in our interviews with manual and non-manual shop stewards and with works managers. But as we explain in the body of the book, we were able to extend the questioning in those interviews by, for instance, differentiating between reactions to changes first, when they were initially proposed and, secondly, when they were implemented. Similarly, we were able to differentiate between consultations about whether changes should be introduced and consultations over the implementation of change. The full interview schedules are available from the ESRC Data Archive at the University of Essex.

- 28 -

		Col./ Code	Skip to
	SECTION SIX		
	MICROTECHNOLOGY AND THE MANUAL WORKFORCE		
	I would like to ask you some questions about the use of the new microelectronics technology as it affects your manual workforce. SHOW CARD J On this card is an explanation of what we include in our use of the phrase.		
56.	INTERVIEWER CHECK Q.1b)		
	ESTABLISHMENT MANUFACTURING (Code 1)	A	Q.57
	ESTABLISHMENT NON-MANUFACTURING (Code 2)	B	Q.60
57.a)	Are you at present using the new microelectronics technology here in any of your products?	(1207)	
	Yes	1	b)
	No	2	Q.58
	INCLUDE IN 'YES' BUYING IN OF COMPLETE COMPONENTS THAT INCORPORATE MICRO-TECHNOLOGY.		
	IF YES AT a)		
	b) In what year did you first start making a product that incorporated microelectronics?		
	WRITE IN: 1 9 _ _	(1208-9)	
	c) Approximately what proportion of your manual workforce are working on products that incorporate microelectronics?		
	WRITE IN: % _ _ _	(1210-2)	
58.a)	Are you at present using the new microelectronics technology in any of your production processes here, including computer controlled plant, machinery or equipment?	(1213)	
	Yes	1	b)
	No	2	See Q.59
	IF YES AT a)		
	b) In what year did you first introduce production plant or processes that incorporated microelectronics?		
	WRITE IN: 1 9 _ _	(1214-5)	
	c) Approximately what proportion of your manual workforce is working on plant or processes that incorporate microelectronics?		
	WRITE IN: % _ _ _	(1216-8)	

- 29 -

			Col./ Code	Skip to
	IF YES AT Q.57a) OR Q.58a) OTHERS SKIP TO Q.61			
	SHOW CARD K			
59.	Are you using microelectronics in the following applications? PROMPT: Any others? UNTIL 'No'.	Yes	No	
		(1219)		
	Design	1	A	Q.61
	Machine control (of individual machines)	2	A	
	Process control (of individual items of process plant)	3	A	
	Centralised machine control (of groups of machines)	4	A	
	Integrated process control (of several stages of processes)	5	A	
	Automated handling (of products, materials or components)	6	A	
	Automated storage	7	A	
	Testing, quality control	8	A	
	IF NON-MANUFACTURING ESTABLISHMENT (CODE B AT Q.56) OTHERS SKIP TO Q.61			
60.a)	Are you at present using the new microelectronics technology in any applications that have directly affected the jobs of manual workers, including the replacement of jobs formerly held by manual workers? INCLUDE COMPUTER CONTROLLED PLANT, MACHINERY OR EQUIPMENT		(1220)	
		Yes	1	b)
	IF YES AT a)	No	2	Q.61
	b) In what year did you first use any application of microelectronics that affected the jobs of manual workers in any way?			
	WRITE IN: 1 9 _ _		(1221-2)	
	c) Approximately what proportion of your manual workforce are now working in jobs directly affected by microelectronics applications?			
	WRITE IN %: _ _ _		(1223-5)	
61.	SHOW CARD L Would you look at this card and tell me which, if any, of these changes you have made during the past three years directly affecting the jobs or working practices of any section or sections of the manual workforce? CODE ALL THAT APPLY			
		Yes	No	
		(1226)		
	A. The introduction of new plant, machinery or equipment, that includes the new microelectronics technology (including computer controlled plant, machinery or equipment)	1	A	
	B. The introduction of new plant, machinery or equipment, not incorporating microelectronics (excluding routine replacement)	2	A	
	C. Substantial changes in work organisation or working practices not involving new plant, machinery or equipment	3	A	
	NONE OF THESE	4		

- 30 -

		Col./ Code	Skip to
·2.	CHECK BWDS		
	FEWER THAN 25 MANUAL EMPLOYEES	A	Q.73
	25 OR MORE MANUAL EMPLOYEES	B	Q.63
	REFER TO ANSWERS TO Q.61	(1227)	
	- IF 'YES' AT A, ASK Q.63-72 ABOUT MOST RECENT MAJOR CHANGE INVOLVING MICROELECTRONICS		
	- IF 'NO' AT A, BUT 'YES' AT B ASK Q.63-72 ABOUT MOST RECENT MAJOR CHANGE NOT INCORPORATING MICROELECTRONICS		
	- IF 'NO' AT A AND B BUT 'YES' AT C, ASK Q.63-72 ABOUT MOST RECENT MAJOR CHANGE NOT INVOLVING NEW PLANT.	(1228)	
	- IF 'NONE OF THESE' SKIP TO Q.73		
·3.	Now I would like to ask you some questions that focus on your most recent major change involving ... READ A, B, OR C AS APPROPRIATE.		
a)	Would you briefly describe what was involved in the change?	(1229)	
	____________________		
	____________________		
	____________________		
b)	How many months ago was the change completed?	(1230)	
	Not yet complete	1	
	Less than 3 months ago	2	
	3 months - less than 6 months	3	
	6 months - less than 9 months	4	
	9 months - less than 12 months	5	
	12 months - less than 18 months	6	
	18 months - less than 2 years	7	
	2 years - less than 3 years	8	
c)	Approximately what proportion of the manual workforce at this establishment were directly affected by the change?		
	WRITE IN: % ☐☐☐	(1231-3)	

- 31 -

		Col./Code	Skip to
64.a)	REFER TO Q.3a)		
	IF SINGLE ESTABLISHMENT (Code 1)	A	Q.65
	IF PART OF LARGER ORGANISATION (Code 2)	B	b)
b)	I now want to distinguish between the decision to introduce the change and the decisions about how the change should be introduced. Firstly, was the decision to make the change taken by management at this establishment or by management at a higher level in the organisation? RECORD ANSWER BELOW IN COL b) RING ONE CODE ONLY		
c)	Secondly, was the decision about how the change should be introduced taken by management at this establishment or management at a higher level in the organisation? RECORD ANSWER BELOW IN COL c) RING ONE CODE ONLY		

	b)	c)
	(1234)	(1235)
Decisions taken at this establishment	1	1
Decision taken at a higher level in this organisation	2	2
Joint decision	3	3
Other answer (SPECIFY) ______________________	4	4

			Col./Code	Skip to
	ALL WITH CHANGE			
65.a)	(IF Q.64 ASKED, BEGIN - 'Apart from management at a higher level in the organisation ...) ... was any organisation or person outside the establishment consulted about how the change should be introduced so far as manual workers were concerned?		(1236)	
		Yes	1	b)
		No	2	Q.66
	IF YES AT a) b) What body(ies) or person(s) were consulted about how the change should be introduced so far as manual workers were concerned? PROBE: Any others? Until 'No'. OBTAIN FULL LIST OF TYPES OF ORGANISATIONS OR INDIVIDUALS CONSULTED, BUT NOT PERSONAL NAMES		(1237)	
	______________________		(1238)	
	______________________		(1239)	

		Col./Code	Skip to
66.	Now I would like to ask about any discussions or consultations between management and workers or their representatives about the introduction of the change or how it was to be implemented.		
	SHOW CARD M Please look at this card. Were discussions or consultations of any of these types held about the introduction of the change or how it was to be implemented? IF YES: Which ones? PROBE: Any others? UNTIL 'No'. RING ALL CODES THAT APPLY	(1240)	
	A Informal discussions with individual manual workers	1	
	B Meetings with groups of manual workers	2	
	C Discussions in established joint consultative committee	3	
	D Discussion in specially constituted committee to consider the change	4	
	E Discussions with union representatives at the establishment	5	
	F Discussions with paid union officials outside the establishment	6	
	None of these	7	
67.	INTERVIEWER CHECK Q.29		
	NO MANUAL UNIONS RECOGNISED AT ESTABLISHMENT (Code A)	A	Q.69
	ONE OR MORE MANUAL UNIONS RECOGNISED AT ESTABLISHMENT (Coded B or C)	B	Q.68
68.	Would you say that the introduction of this change was ... READ OUT AND RING ONE CODE ONLY ...	(1241)	
	... negotiated with union representatives and dependent on their agreement,	1	
	or, discussed with union representatives in a way which took their views into account but left management free to make the decisions,	2	
	or, not discussed with union representatives?	3	
	ALL WITH CHANGE		
69.a)	In general, did the change we have been discussing result in the earnings of the manual workers directly affected being ... READ OUT ...	(1242)	
	... increased,	1	b)
	or decreased?	2	Q.70
	(Neither - no change)	3	Q.70
	IF INCREASED AT a)		
	b) Was the increase the result of ... READ OUT AND RECORD EACH IN TURN.	Yes / No	(1243)
	Regrading or upgrading?	1 / A	
	Higher or new bonuses?	2 / B	
	More paid overtime?	3 / C	
	Higher rates agreed as part of the agreement to accept the change?	4 / D	
	Anything else? (SPECIFY) ______________________	5 / E	
70.	Did the change we have been discussing result in manning or staffing levels in the sections of manual workers directly affected being ... READ OUT ...	(1244)	
	... increased,	1	
	or decreased?	2	
	(Neither - no change)	3	

- 33 -

		More	Less	No change	Can't say	Col./Code	Skip to
71.	Would you say that most of the manual workers affected by the change ... READ OUT ...						
	Have more or less responsibility	1	2	3	4	(1245)	
	Have to work at a more skilled or a less skilled level	1	2	3	4	(1246)	
	Have more control or less control over their pace of work	1	2	3	4	(1247)	
	Have more or less control over how they do their jobs	1	2	3	4	(1248)	
	Are subject to more or less supervision	1	2	3	4	(1249)	
	Have more interesting or less interesting jobs	1	2	3	4	(1250)	
		Wider	Narr-ower				
	Have a wider or narrower range of tasks to do	1	2	3	4	(1251)	

72. Finally, on this topic, I would like to ask about the reactions of different groups of workers or their representatives when you were bringing in the change.

a) SHOW CARD N First, so as far as foremen and supervisors of manual workers were concerned, what would you say their reactions were? Please choose a phrase from this card. RECORD ANSWER IN COL a) BELOW

b) And so far as the workers directly affected by the change ...? RECORD ANSWER IN COL b) BELOW.

IF MANUAL UNIONS RECOGNISED (CHECK Q.27a) OTHERS SKIP TO Q.73

c) And so far as the shop stewards or manual worker representatives here ...? RECORD ANSWER IN COL c) BELOW

d) And so far as paid union officials outside the establishment ...? RECORD ANSWER IN COL d) BELOW

	a)	b)	c)	d)
	(1252)	(1253)	(1254)	(1255)
Strongly resistant	1	1	1	1
Slightly resistant	2	2	2	2
Slightly in favour	3	3	3	3
Strongly in favour	4	4	4	4
Does not apply	-	-	5	5

Col./Code	Skip to
(1256-80)	Blank
(1301-4)	Repeat
(1305-6)	CARD 13

- 48 -

SECTION TEN

MICROTECHNOLOGY AND THE NON-MANUAL WORKFORCE

		Col./Code	Skip to

105. I'd now like to talk about any computing arrangements you have at this establishment.

SHOW CARD O Which, if any, of these computing facilities are in use at this establishment?

CODE ALL THAT APPLY

	Yes	No
	(1507)	
A main frame computer on this site	1	A
A link to a computer sited at another establishment in the organisation	2	A
On site mini computer	3	A
On site micro computer	4	A
A link to an external computer belonging to another organisation	5	A
The use of bureaux services	6	A
None of these	7	

106. Do you have any word processing equipment at this establishment?

	Col./Code	Skip to
	(1508)	
Yes	1	b)
No	2	Q.107

IF HAVE WORD PROCESSING EQUIPMENT

a) How many separate stations/terminals with visual display units and keyboard facilities, do you have here?

WRITE IN NO: [|] (1509-10)

b) In what year did you first introduce word processing equipment at this establishment?

WRITE IN: 19 [|] (1511-2)

107. Now, I would like to ask you some questions about changes affecting office workers in this establishment. By this we mean clerical, secretarial, administrative and typing staff. (1513-25) Blank

a) SHOW CARD P Would you look at this card and tell me which, if any, of these changes you have made during the past three years directly affecting the jobs or working practices of clerical, or secretarial or administrative or typing staff? CODE ALL THAT APPLY
PROBE: Any others? UNTIL 'No'.

	Yes	No
	(1526)	
... A. The introduction of word processing or computer applications	1	A
... B. The introduction of new machinery or equipment, not involving word processor or computer applications (excluding routine replacement)	2	A
... C. Substantial changes in work organisation or working practices not involving new machinery or equipment	3	A
None of these	4	

- 49 -

		Col./ Code	Skip to
108.	CHECK BWDS		
	FEWER THAN 25 NON MANUAL EMPLOYEES	A	Q.120
	25 OR MORE NON MANUAL EMPLOYEES	B	Q.109
	REFER TO ANSWERS TO Q.107	(1527)	
	- IF 'YES' AT A, ASK Q.109-119 ABOUT MOST RECENT MAJOR CHANGE INVOLVING WORD PROCESSORS OR COMPUTER APPLICATIONS		
	- IF 'NO' AT A, BUT 'YES' AT B ASK Q.109-119 ABOUT MOST RECENT MAJOR CHANGE NOT INCORPORATING WORD PROCESSOR OR COMPUTER APPLICATIONS		
	- IF 'NO' AT A AND B, BUT 'YES' AT C, ASK Q.109-119 ABOUT MOST RECENT MAJOR CHANGE NOT INVOLVING NEW PLANT	(1528)	
	- IF 'NO' AT A, B AND C, SKIP TO Q.120		
109.	Now I would like to ask you some questions that focus on your most recent major change involving ... READ A, B OR C AS APPROPRIATE.		
a)	Would you briefly describe what was involved in the change?	(1529)	
	______________________________		
	______________________________		
	______________________________	(1530)	
b)	How many months ago was this change completed?		
	Not yet complete	1	
	Less than 3 months ago	2	
	3 months - less than 6	3	
	6 months - less than 9	4	
	9 months - less than 12	5	
	12 months - less than 18	6	
	18 months - less than 24	7	
	2 years - less than 3 years	8	
c)	Approximately what percentage of the office workers at this establishment were directly affected by the change? WRITE IN %: [ ][ ][ ]	(1531-3)	
110.	REFER TO Q.3a) IF SINGLE ESTABLISHMENT (CODE 1)	A	Q.112
	IF PART OF LARGER ORGANISATION (CODE 2)	B	Q.111
111.	I now want to distinguish between the decision to introduce the change and the decisions about how the change should be introduced.		
a)	Firstly, was the decision to make the change taken by management at this establishment or by management at a higher level in the organisation? RECORD ANSWER BELOW IN COL. a) RING ONE CODE ONLY		
b)	Secondly, was the decision about how the change should be introduced taken by management at this establishment or management at a higher level in the organisation? RECORD ANSWER BELOW IN COL. b). RING ONE CODE ONLY		

	a)	b)
Decisions taken at this establishment	1	1
Decision taken at a higher level in this organisation	2	2
Joint decision	3	3
Other answer (SPECIFY) ______________________________	4	4
______________________________		
	(1534)	(1535)

- 50 -

		Col./ Code	Skip to
	ASK ALL WITH CHANGE		
112.a)	(IF Q.111 ASKED, BEGIN - 'Apart from management at a higher level in the organisation ...)		
	... was any organisation or person outside the establishment consulted about how the change was to be introduced as far as office workers were concerned?	(1536)	
	Yes	1	b)
	No	2	Q.113
	IF YES AT a)	(1537)	
	b) What organisations or person(s) were consulted about how the change was to be introduced so far as office workers were concerned? PROBE: Any others? Until 'No'.		
	OBTAIN FULL LIST OF TYPES OF ORGANISATIONS OR INDIVIDUALS CONSULTED, BUT NOT PERSONAL NAMES	(1538)	
	______		
	______		
	______	(1539)	
	______		
113.	Now I would like to ask about any discussions or consultations between management and workers or their representatives about the introduction of the change or how it was to be implemented?	(1540)	
	SHOW CARD Q Looking at this card, were discussions or consultations of any of these types held about the implementation of the change or how it was to be implemented? CODE ALL THAT APPLY		
	A. Informal discussions with individual office workers	1	
	B. Meetings with groups of office workers	2	
	C. Discussions in established joint consultative committee	3	
	D. Discussion in specially constituted committee to consider the change	4	
	E. Discussions with union or staff association representatives inside the establishment	5	
	F. Discussions with paid union/assoc officials outside the establishment	6	
	None of these	7	
114.	INTERVIEWER CHECK Q.78		
	NO NON MANUAL UNIONS/ASSOCIATIONS RECOGNISED AT ESTABLISHMENT (Code A)	A	Q.116
	ONE OR MORE NON MANUAL UNIONS/ASSOCIATIONS RECOGNISED AT ESTABLISHMENT (Coded B or C)	B	Q.115
115.	Would you say that the introduction of this change was ... READ OUT AND RING ONE CODE ONLY ...	(1541)	
	... negotiated with union or staff association representatives and dependent on their agreement	1	
	or, discussed with union or staff association representatives in a way which took their views into account but left management free to make the decisions	2	
	or, not discussed with union or staff association representatives?	3	

- 51 -

		Col./ Code	Skip to
	ALL WITH CHANGE		
116.a)	In general, did the change we have been discussing result in the earnings of office workers directly affected being ... READ OUT ...	(1542)	
	... increased	1	b)
	or decreased?	2	Q.117
	(Neither - no change)	3	Q.117
	IF INCREASED AT a)		
	b) Was the increase the result of ... READ OUT AND RECORD EACH IN TURN CODE ALL THAT APPLY	Yes No	(1543)
	Regrading or upgrading?	1 A	
	Higher bonuses or payments by results?	2 B	
	More paid overtime?	3 C	
	Higher rates agreed as part of the agreement to accept the change?	4 D	
	Anything else? (SPECIFY) ______	5 E	
117.	Did the change we have been discussing result in staffing or manning levels in the section(s) of office workers affected being ... READ OUT ...	(1544)	
	... increased	1	
	or decreased?	2	
	(Neither - no change)	3	

118. Would you say that most of the office workers affected by the change ... READ OUT ...

	More	Less	No change	Can't say	
Have more or less responsibility	1	2	3	4	(1545)
Have to work at a more skilled or a less skilled level	1	2	3	4	(1546)
Have more control or less control over their pace of work	1	2	3	4	(1547)
Have more or less control over how the job is done	1	2	3	4	(1548)
Are subject to more or less supervision	1	2	3	4	(1549)
Have more interesting or less interesting jobs	1	2	3	4	(1550)
	Wider	Narr-ower			
Have a wider or narrower range or tasks to do	1	2	3	4	(1551)

- 52 -

		Col./Code	Skip to

119. Finally, on this topic I would like to ask about the reactions of different groups of workers or their representatives when you were bringing in the change.

a) First, so far as supervisors of the office workers were concerned, what would you say their reactions were? SHOW CARD N Please choose a phrase from this card. RECORD ANSWER IN COL a) BELOW

b) And so far as the workers directly affected by the change ...? RECORD ANSWER IN COL b) BELOW

IF NON MANUAL UNIONS/ASSOCIATIONS RECOGNISED (CHECK Q.78)

c) And so far as the non manual representatives or stewards here? RECORD ANSWER IN COL c) BELOW

d) And so far as paid union or association officials outside the establishment ...? RECORD ANSWER IN COL d) BELOW

	a)	b)	c)	d)
	(1552)	(1553)	(1554)	(1555)
Strongly resistant	1	1	1	1
Slightly resistant	2	2	2	2
Slightly in favour	3	3	3	3
Strongly in favour	4	4	4	4
Does not apply	-	-	5	5

Col./Code	Skip to
(1556-80)	Blank
(1601-4) (1605-6)	Repeat CARD 16

Appendix B

Summary of the Technical Aspects of the Survey

A fuller account of the technical aspects of WIRS 84 appears in our companion volume(1). Here, we simply summarise the main features of the Technical Appendix in that book.

The sampling frame and the sample
The sample design for the 1984 survey closely followed that developed for the 1980 survey. The sampling frame was the Department of Employment's 1981 *Census of Employment*; for the 1980 survey it was the Census conducted in 1977. As in 1980, all census units recorded as having 24 or fewer employees were excluded, as were units falling within Agriculture, Forestry and Fishing (Division 0) of the *Standard Industrial Classification (1980)*. Otherwise all sectors of civil employment in England, Scotland and Wales were included in the sampling universe – public and private sector, manufacturing and service industries. In 1984, as in 1980, differential sampling fractions were used for the selection of units of different sizes (in terms of numbers of full-time and part-time employees: 25 to 49, 50 to 99 and so on).

A census unit is in most cases a number of employees working at the same address who are paid from the same location by the same employer. The requirement of the survey design was for a sample of *establishments*, that is, of individual places of employment at a single address and covering all the employees at the address of the identified employer. In general there is a sufficient degree of one-for-one correspondence between census units and establishments for the census to serve as a viable sampling frame for the survey series. But in some cases census units refer to more than one establishment, and in those cases we devised a procedure for identifying an appropriate establishment.

At the time of the design of the 1984 sample, the 1981 *Census of Employment* files contained just over 135,000 units recorded as having 25 or more employees. From these files a stratified random sample totalling 3640 units was drawn.

Differential sampling fractions were used for selection within each of seven size bands, the size bands being determined by the number of full and part-time employees recorded as being present. Within each size band the census units were initially stratified by: the proportion of *male* employees, within the proportion of *full-time* employees, within the *Activities* of the *Standard Industrial Classification (1980)*. From the resultant list, samples were selected by marking off at intervals from a randomly selected starting-point, the list being treated as circular. The numbers of census units, sampling fractions and subsample sizes are given in Table A.1.

Table A.1 Sampling fractions and numbers of census units drawn

Numbers

	No. of units in census 1984	Sampling fraction 1984	Sample selected 1984
No. of employees recorded at census unit:		(1 in . . .)	
25-49	70,000	92	760
50-99	33,288	51	650
100-199	17,625	28	620
200-499	9,880	16	600
500-999	2,796	6	500
1000-1999	1,169	3.3	360
2000+	484	3.3	150
Total	135,242	37	3,640

Before the start of fieldwork, 43 census units were eliminated from the sample, because they were outside the scope of the survey (19); or because approaches had been made either to those addresses or to addresses in the same organisation during the course of piloting and development (19); or because the addresses had been printed out incompletely or obscurely or duplicated (5). Finally, all addresses in the coal-mining industry were withdrawn at the start of the fieldwork because of the continuing industrial dispute in that sector and the consequent unlikelihood of successful interviewing.

Questionnaire development and fieldwork

A pilot survey of 56 establishments was conducted during December 1983 and early January 1984. The interviewing for the main survey was carried out by 117 interviewers after initial contact had been made by letter from the Department of Employment. Five of the interviewers were SCPR employees, being members of the field staff. Apart from three relatively recently recruited interviewers, the remaining 109 were experienced interviewers from SCPR's regular interviewing panel. All had been trained by SCPR's field staff and 46 of the 117 had also worked as interviewers on the 1980 survey. Before fieldwork began, seven briefing conferences were conducted jointly by members of the research team from the Department of Employment, the Policy Studies Institute and SCPR. One hundred and five interviewers attended this series of conferences during early March 1984. The remaining twelve were briefed at four supplementary conferences by SCPR research and field staff during April and May. Fieldwork began immediately after the main interviewer briefings and continued until October 1984.

Fieldwork quality control was carried out by a postal check on a random subsample of establishments after the fieldwork period was over. In addition to these postal checks the first batch of questionnaires received in field offices from each interviewer in the 1984 survey was subject to intensive checking by clerical staff and, if necessary, remedial action was taken.

Overall response

The outcome of the sampling, initial approach and fieldwork operations may be judged from the summary statistics in Table A.2. Ineligible or out-of-scope addresses fell into three main groups: those which were found to have closed down between the taking of the Census (1981) and interviewing (1984), of which there were 180, those of establishments which were found to have fewer than 25 employees, of which there were 178; and those which were found to be vacant or demolished premises or where the establishment had moved, leaving no trace of its new whereabouts, of which there were 42.

Table A.2 Summary of fieldwork response

Numbers

	Addresses	
	1980	1984
Initial sample	3307	3154
Resampled units[a]	25	55
Total sample	3332	3209
Less:		
Withdrawn at sampling stage	205	135
Ineligible/out-of-scope	376	449
Non-productive addresses	686	606
Interviews achieved	2040	2019

[a] Units resampled from aggregate census returns.

Non-productive addresses fell into three main groups: those for which a refusal was received at the Department of Employment in response to the original letter sent out, of which there were 220 cases; those for which a refusal was received by the SCPR interviewer or at SCPR offices, of which there were 316 or just over 13 per cent of all addresses passed to interviewers; and those at which no effective contact was made (41 cases) or at which questionnaires were completed but could not be used (29 cases).

The overall response rate, judged by the completion of at least a satisfactory management interview, was 77 per cent. As a fieldwork operation, setting aside those addresses that were not issued to interviewers, the response rate achieved was 84 per cent. The overall response rate was analysed by region, industrial activity and establishment size. In regional terms, variations in response, ranged from 74 per cent in the West Midlands to 82 per cent in the North. Response rates for different industrial sectors ranged from 66 per cent in 'Other manufacturing' to 91 per cent in 'Energy and water supply' (which, as noted earlier, excluded coal mining). In the main, however, the response rates fell within the range 72 per cent to 85 per cent. Response rates for different sizes of establishment ranged from 71 per cent in the smallest workplaces (25–49 employees) to 85 per cent in workplaces employing between 1000 and 1999 people.

Response among worker representatives and works managers

Within the 2019 establishments at which interviews were obtained with management respondents in 1984, manual negotiating groups were identified in 1327 cases (1334 in 1980) and non-manual in 1367 cases (1255 in 1980). In a small number of establishments (not included in the figure above) negotiating groups were not positively identified, although it was apparent from subsequent answers that management at the establishment did recognise one or more trade unions for negotiating purposes. Typically, the reason for such ambiguity was that negotiations took place at a higher level in the organisation. The numbers of such establishments were 78 in respect of manual recognition (31 in 1980) and 31 in respect of non-manual recognition (27 in 1980). In a further 174 cases there was no manual trade union representative at the sampled establishment and in 205 non-manual cases an equivalent situation existed. In the remaining cases — those where appropriate representatives existed at the sampled establishment — interviewers were required to seek an interview with the representative of the primary negotiating group for manual employees and, separately, with the representative of the primary negotiating group for non-manual employees.

In all, 910 interviews with primary manual representatives were achieved and 949 with primary non-manual representatives. At those establishments where worker representatives were present, the overall response rates were 79 per cent for manual representatives and 82 per cent for non-manual representatives. The major reason for failing to obtain an interview was the refusal to grant permission by management, but there was also an increase in the rate of refusals among worker representatives themselves, compared with our experience in 1980.

Interviewers were also required to conduct additional interviews with the works manager at those manufacturing plants where the main management respondent was a specialist in industrial, employee or staff relations or in personnel. Such specialists were present in 365 of the 561 manufacturing plants in the 1984 sample, but in 15 of these there was no works manager present at the establishment. Of the remaining 350 establishments, works managers were successfully interviewed in 259 cases, a response rate of 74 per cent. The most common reason for non-response was a refusal by the main management respondent. However, the dataset includes a total of 276 works manager interviews because an additional 17 interviews were carried out at manufacturing establishments where the main management respondent had *not* been coded as a personnel or industrial relations specialist. Strictly, this is a matter of interviewer error, but may also reflect the situation at the establishment. The research team examined these questionnaires during the coding and editing stages and decided to include them in the dataset, taking into account the element of arbitrariness in the definition of the circumstances in which these interviews were required.

Coding and editing of the data

Coding and editing of the questionnaires was carried out between June and December 1984. Rigorous structure checking of all questionnaires was undertaken prior to general coding and editing. Particular attention was paid to the *Basic Workforce Data Sheets (BWDS).* The absence of these or obscurities or inconsistencies in their completion were the most common reasons for returning questionnaires to interviewers.

Computer editing of the data was carried out in two stages. The first edit program consisted of a rigorous check of ranges and filters and questionnaire structure, while the second comprised a number of logic checks, checks on extreme values and on relatively complex relationships between different sections of the questionnaires. The complete data file incorporating weighting was handed over to the research team on 16 April 1985. Further detailed work on the file was also carried out in order to provide an anonymised version for the ESRC Data Archive at Essex University. The anonymised tape was lodged at the Archive at the beginning of August 1985.

Weighting of the data

All the results presented in the main text of this report, unless otherwise specified, were adjusted by weighting factors derived from two separate stages of calculation. The first stage compensated for the inequalities of selection that were introduced by the differential sampling of census units according to the number of their employees.[2] This first stage of weighting is imperative, otherwise the results from each size stratum simply cannot be added together to provide a meaningful aggregate. The second, and additional, stage of weighting was applied to adjust for the observed under-representation of small establishments in the distribution resulting from the first stage. Ideally, we would have matched the size distribution of the sample to the 1984 *Census of Employment* size distribution; unfortunately this was not available during the course of our analysis. Further details are given in a note available from the ESRC Data Archive. The additional factors derived from the second stage of weighting are incorporated into a single set of weights, representing both stages, which are referred to as the *second weighting scheme*. It is this that we have applied throughout.

The panel sample

It was decided to include in the 1984 survey an experimental panel consisting of establishments included in the 1980 survey. We have not used that panel in the present volume because we have been concerned with the new information collected in WIRS 84 rather than with the measurement of change as revealed by answers to the core questions asked in both 1980 and 1984. Further details of the panel appear in our companion volume.

Sampling errors

Sampling errors for the main cross-sectional sample are somewhat larger than for a simple random sample of equivalent size, owing to the disproportionate sampling of larger establishments. Sampling errors were calculated for a number of variables in the 1980 survey and ranged from 0.4 per cent to 1.4 per cent. Given the very similar design and sample size of the two surveys, sampling errors for the 1984 survey are unlikely to be very different. Work is proceeding and will be reported via the ESRC Data Archive.

Notes and references

1 Neil Millward and Mark Stevens, *British Workplace Industrial Relations 1980–1984*, Gower, 1986. The material on the 1984 survey contained in that appendix is largely based upon technical reports written by Colin Airey of Social and Community Planning Research. Denise Lievesley of the Survey Methods Centre provided material on sampling and on the weighting of the data. The full documents are available from the ESRC Data Archive.

2 These weighting factors are also contained in the data records held at the ESRC Data Archive.

Index